THE ROAD TO THE
STARS

THE ROAD TO THE
STARS

★

Iain Nicolson

William Morrow & Company, Inc.

New York 1978

Contents

Acknowledgements

It is impossible to acknowledge fully the sources of many of the ideas which appear in a book of this nature, as these have come together over many years of reading and discussions with colleagues and students. To these unnamed sources, I am grateful. A considerable debt of gratitude is due to Andrew Farmer for his excellent colour illustrations and to the British Interplanetary Society, in particular to Drs A. R. Martin and Alan Bond, for their substantial assistance in the preparation of the Project Daedalus artwork. I should also like to thank NASA for a liberal supply of information and illustrative material, and in this context especial thanks are due to Frank E. Bristow (Jet Propulsion Laboratory), John E. McLeaish (Johnson Space Center), John B. Taylor (Marshall Space Flight Center), and Paul T. Bohn (Lewis Research Center). I am also grateful to the Novosti Press Agency and to the Hale observatories for further illustrations. My sincere thanks go to Paul Barnett of David and Charles who first suggested the idea of this book to me and who has given first-rate support throughout its preparation, to my wife, Jean, for reading and correcting the typescript, and to my family for putting up with me over the past year!

None of the aforementioned can be held responsible for any errors or inadequacies in the book; these must be laid at my door.

Iain Nicolson
February 1978

Preface

The whole idea of travel to the stars is often dismissed as 'impossible', 'idle speculation', or 'so far beyond the realms of foreseeable technology as not to be worth serious consideration'. I do not share this view. Although we do not yet possess the means to undertake meaningful interstellar exploration, the *fact* is that there already exist space-probes which are moving out of the Solar System never to return; the first journeys into the interstellar void have begun. Admittedly these craft would take hundreds of thousands of years to reach even the nearest stars, but there are propulsion techniques presently being discussed which—without stretching existing scientific knowledge and technology *too* far—may permit us to accomplish such journeys in periods measured in decades rather than millennia. Given the immense funding required for research and development there is good reason to suppose that a starship of this nature could be constructed during the twenty-first century.

This book sets out the view that, provided mankind survives the crises of the next fifty to a hundred years, interstellar travel will become technically feasible. Whether or not it appears to be essential, desirable, or even undesirable, that it will eventually occur seems as inevitable as the process of evolution itself.

The book is divided into three parts. Part I is concerned with the reasons why we may set off along the road to the stars. Part II—the major section—examines the basic principles, progress to date, the immediate future, problems of interstellar flight, and the means by which these journeys may be accomplished. Part III examines some of the likely effects of interstellar travel on individuals and on the human species as a whole. Much of what is said will prove to be very wide of the mark, but I should be surprised indeed if the basic ideas are *all* wrong!

Iain Nicolson
February 1978

Introduction:
out of the cradle

The best current estimates indicate that the Earth was formed 4.6 billion years ago.* Life in the form of bacteria emerged over 3 billion years in the past, followed by the appearance of green and blue-green algae some 500 million years later. The evolution of life was at first a rather slow process, taking place in seas, lakes, pools and inlets, where single-celled creatures dominated for well over two billion years; the complex variety of life which we see today developed only in the last 600 million years, and it was a mere 400 million years ago, or thereabouts, that the first amphibious creatures crawled out of their watery cradle onto the land.

Since that time there has been rapid development in the variety and complexity of species. The reptiles first came to grips with the problems of living on land, and from these evolved the giant dinosaurs which made their appearance about 200 million years ago and which dominated the land masses for some 130 million years. Mammals and birds emerged shortly after the dinosaurs, but it was only with the demise of the latter—presumably due to an inability to adapt to a changing environment—that mammals began truly to flourish. Although Man-like apes have existed for some fifteen million years, Man's true ancestor, a creature which walked upright on two legs, appeared as recently as two to four million years ago.

Early Man was a hunter, forced to struggle for his very existence in a harsh and competitive world. His greatest asset was his brain: superior to that of any other animal (apart, perhaps, from whales and dolphins), it endowed Man with the ability to think ahead, to work through and around problems. Despite his inadequate physical armoury he has been able, in a period of time which is extremely short by geological standards, to dominate his world and all other terrestrial species.

It is less than 10,000 years since Neolithic men began to live in villages, to cultivate plants for food, and to domesticate animals such as sheep and goats. Man the nomadic hunter began to change to a community-dweller and farmer. The earliest records of civilizations, dating

*Throughout this book the US billion—one thousand million—is used in preference to the European billion—one million million.

from rather more than five thousand years ago, relate to the inhabitants of the region of the rivers Tigris, Euphrates and Nile, and already by this time some of mankind had risen a little above the raw struggle for the necessities to sustain life and were able to contemplate the wonders of the Universe around them. Civilization spread across Europe and Asia, developing through city states, empires and nation states to the complex of power blocs and alliances which we see today; high-speed communications and transport systems seem now to be leading us towards one global culture.

Some 250 years ago there began the Industrial Revolution, the application of technology to the production of machines which greatly enhanced the speed at which goods could be produced, and since the invention of the steam engine in 1769 Man has had controllable power at his fingertips. A new industrial revolution began in the twentieth century with the advent of automation, the development of electronics and the emergence of the computer, which gave Man the means to handle vast quantities of data.

Throughout the period of recorded history, Man's inquisitive nature has been apparent. His desire to know and understand his world has been the driving force behind the development of his philosophies and his science, and this 'need to know' has radically altered Man's view both of the Universe and of his own place in it. We have moved from the Earth-centred Universe of the ancient Greeks to the Sun-centred Universe of Copernicus, and on to a concept of a Universe in which neither the Sun nor our Galaxy occupies any central position, a Universe which may not even *have* a centre. As our intellectual horizons have expanded so have the physical boundaries of Man's domain. Man's explorations have taken him to the remotest corners of the globe, to the highest mountain peaks and to the oceans' depths. This 'need to explore' is surely deep-seated in the nature of mankind, and is not motivated by the prospect of obvious or immediate economic gain. The profit which might accrue from the fifteenth-century ocean voyages or the earlier Nordic voyages could scarcely have been obvious at the time, but the discovery and development of the New World of the Americas has had a crucial influence on the history of mankind. The launching of the first satellites and spaceprobes, the first manned space-flights, the journey of Apollo to the Moon, none of these things were motivated by any consideration of material gain—although that will come and, indeed, the space environment is already big business. It is mankind's need to explore, and his need of 'challenge', which have been the driving force behind his explorations, both physical and intellectual.

The need of 'challenge' is readily apparent in the developed industrial nations, where we are so far removed from the struggle just to stay alive that we have considerable leisure time. We find it necessary to invent our own challenges, in climbing mountains, sailing the oceans, diving to the ocean depths or soaring to great heights, pushing ourselves to the limits of our capabilities, and beyond. Space offers the new challenge, and the tantalizing new frontier for physical exploration.

The whirlwind pace of mankind's progress has been sustained at a cost. With the rapid advance of technology Man's dominance over the natural world has become so great that a severe imbalance has arisen; he is rapidly and radically altering his environment by both accident and design to the extent where serious global consequences may ensue. We are at a crucial point in the history of the world. The actions of mankind in the next few decades will very likely decide whether or not *Homo sapiens* is destined to go forward and evolve further or to subside into oblivion along with the host of previous evolutionary 'mistakes'. Whereas the dinosaurs vanished after a long period of domination because they could not adapt to a changing environment, the human race may perish as a direct consequence of its own effect on the environment.

The great challenge of our times is the survival of the human species, or, at least, the survival of civilization as we know it. The problem which is most frequently debated at present is the so-called 'energy crisis', which arises because the fossil fuel resources of this planet are running out and cannot be replaced. The same problem arises for readily accessible high-quality mineral resources. These are, of course, serious problems made much worse by the fact that our demands for energy and natural resources are increasing exponentially while the available resources are decreasing exponentially. Our known reserves of fossil fuels, notably coal, are sufficient to sustain our *present* energy requirements for several centuries, but, with escalating demand, the timescale becomes greatly shortened, and there is little doubt that we are heading for a major crisis by the early part of the twenty-first century. 'Solutions' to the energy crisis which have been proposed are many and varied, ranging from a return to small self-sufficient low-technology communities to the wholesale development of nuclear fission power plants with all the lethal radioactive waste which they generate. The ideal solution would seem to be the development of fusion reactors which—operating by a similar process to that which powers the Sun— would produce virtually unlimited power from the 'heavy' hydrogen contained in seawater; despite decades of research, such a development is certainly not to be expected in the immediate future, while the problem itself is a part of the present.

Serious as the energy problem may be, it is not of itself the major problem facing mankind. There surely can be no doubt that the problem that lies at the root of mankind's crisis is overpopulation. Unless and until we can halt the disastrous upward spiral of population and stabilize human numbers on Earth at a level preferably significantly below the present figure, there can be little hope for the future. Not only is the population growing at an alarming rate, but it is growing in such a way as to increase even further the proportion of 'have nots' to 'haves' in this world. United Nations statistics* published in 1977 indicate that, while the population growth rate in the developed nations (or, as Dr Paul Ehrlich, author of *The Population Bomb*, would put it, the *overdeveloped* nations) is currently about 0.9% per annum and dropping, the growth rate in the less developed nations, where standards of living are in any case intolerably low, is about 2.3%. As a result of this, the proportion of humanity possessing a high standard of living dropped from about one third in 1950 to less than a quarter in 1970, while the predictions for the year 2000 indicate that this decline will continue. Increased global communication and awareness leads inevitably to a rise in human expectations and this, coupled with an increase in the proportion of less privileged human beings, must exacerbate the crisis. The already rich nations may (although one doubts it) come to the conclusion that they must ease up on their energy-intensive industrial activities for the sake of the environment of the Earth, but I would be surprised if the rising tide of underdeveloped humanity were greatly impressed by this. We cannot, it seems, live without industry, technology and escalating energy usage, yet these very processes are pitching the Earth into crisis. Can this circle be broken?

It has been suggested that spaceflight could provide the means to solve the population problem, simply by transporting people to other planets. Such a suggestion is absurd. Within the Solar System there are perhaps two planets which could be modified to make habitats for Man, but the technology required is still well beyond our present capabilities. The problem is an immediate one. The world population is *already* too large, being about 4 billion in 1975, compared to only 2.5 billion in 1950. The total world population is growing at 2% per annum, and this implies that it will double every 35 years, reaching about six billion by the year 2000. For the planetary colonization 'solution' to work, something approaching the entire present population of the Earth would have to be transported within 35 years. Venus is about the same size as the Earth, but Mars is appreciably smaller, so that within, say, fifty years of the commencement of the operation (assuming it started tomorrow), if the human population were allowed to continue to grow at its present rate, the capacity of the planets of the Solar System to support the swelling tide of humanity would be exhausted, and we would then have to start transporting people at a frantic rate to other planetary systems.

The absurdity of such an approach is highlighted by considering the entire Galaxy. There are about one hundred billion stars in our Galaxy and it is thought that a considerable proportion of these may have planetary systems, although only a small fraction of these will have conditions remotely suitable for *human* life. But let us be optimistic and

*World Population Prospects as Assessed in 1973, United Nations, New York, 1977.

suppose that every star has one Earth-type planet available for colonization. The expanding mass of human protoplasm, at its present growth rate, would inhabit one hundred billion planets in less than 1300 years! Furthermore, since the Galaxy is about thirty thousand parsecs (100,000 light years) in diameter, this colonization could be achieved only by faster-than-light travel.

The solution of the population problem rests fairly and squarely with the present generation of Earth inhabitants, and that solution must be found and achieved by our own efforts within the next few decades. If we do not limit population by our own efforts, then the forces that will do so—war, famine, disease—will achieve the reduction at the cost of inestimable human suffering.

Having said that the colonization of planets cannot solve the population crisis, I still believe that Man's expansion out into the space environment will be a major factor in the improvement of the quality of life on Earth, and will relieve population pressures of a rather different kind. Industrial processes require an energy source, and sources of raw materials; the processes inevitably produce waste products which pollute the terrestrial environment in a wide variety of ways, from the waste heat radiated into the atmosphere to the toxic wastes which find their way inexorably into our own food chains. A natural ecosystem like the Earth can remain in equilibrium only if the demands which we make upon it are kept within limits, limits which we are already exceeding. In space, energy is available directly from the Sun, minerals may be obtained more economically from the Moon or from the asteroids than from the Earth, and waste products are dispelled into the vastness of the Universe. The place for an energy-hungry, growth-orientated society is in space, not on the surface of a planet like the Earth. The exploitation of the space environment, what G. Harry Stine has called the Third Industrial Revolution, allows economic growth to continue *without* further harming the natural environment of our home planet.

The quest for energy and resources will take mankind to a greater and greater extent into the space around us, and as we become more at home in this expanded environment, so the human population in space will grow. Given a stabilized human population on Earth—as we have seen, preferably a population reduced somewhat from its present level— a considerable measure of control, both voluntary and otherwise, will be essential to maintain a balanced world environment, and this will place physical and political restrictions on the individual. In many ways the Earth would be a much pleasanter planet on which to live, and for most people the loss of some of the freedom to carry out actions without worrying unduly about the consequences would be of little significance. But for those who feel stifled and inhibited by this kind of existence, the road to the stars may offer a way out, and would be an essential safety valve to the terrestrial population.

Powerful arguments for a closely monitored and restricted state of global equilibrium are put forward by the Club of Rome and are presented in *The Limits to Growth*, edited by Meadows *et al.** In such a world state, population and capital would be essentially stable 'with the forces tending to increase them or decrease them in a carefully controlled balance'. In such a society no new technological advance could be adopted until the physical and social side-effects had been established, until the social changes necessary for the development to be implemented had been assessed, and—if the development removed some particular natural restriction on growth—until it had been established what new limit the growing system would meet next. The essence of the argument is that it is better to learn to live with limits than to fight against them.

Such an argument has considerable validity if applied to a wholly closed system. By moving into the space environment, we would be opening up the system and removing its natural limits. Technological advance and economic growth would still be possible for the enhanced human community without harming the environment of the Earth. We could husband the resources of the Earth, expand mankind's horizons and its domain, yet still improve the standard of living of the individual.

As mankind becomes more at home in its enlarged space environment, so the opportunity will arise for separate colonies and communities to diversify and develop cultures different from that of the Earth; the humanization of space may allow us to reverse the trend towards a single bland global culture which is evident on Earth at the moment. Expanding Man's domain will satisfy that powerful inner drive for exploration and will allow communities the freedom to develop along differing lines; it will even open up the possibility of a modest increase in the total human population without overwhelming the Earth's slender resources. It will allow the development of new technologies while preserving the Earth from the undesirable consequences of such development.

Above all we need frontiers, a challenge, a sense of purpose. Perhaps one day the human species will lose its aggression and its questing nature, and will be content to settle for a static, stable existence. For myself, I doubt it, for in so doing we would forfeit those characteristics which are the essence of humanity. If we surmount the global crisis into which we have already entered then I am convinced that we shall move out of the terrestrial cradle into the wider world of the Solar System, and, having expanded our sphere of influence so far, I cannot believe that mankind would call a halt there. Long before the Solar System has yielded all its secrets, the human race will have taken the first steps along the road to the stars and will have launched itself forth into the interstellar void to meet whatever fate has in store for us 'out there'.

*Universe Books, 1972.

No matter what rationalization is put forward to obtain the funding for the exploration and exploitation of space, the fundamental reason why Man will eventually travel to the stars—whether in person or by remote probe—is to respond to the challenge and to satisfy his curiosity, to answer long-standing burning questions—what is the nature of space and time?; are we alone in the Universe?

There is a whole Universe to explore.

PART I

The why of it all

1 A Universe to Explore

Where do we stand in relation to the Universe? We live on the Earth, a small body travelling around the Sun, a typical star. Compared to the Sun, the Earth is relatively dense, being made up largely of heavier elements: metals, such as iron and nickel, and silicates, the rock-forming materials. It has a solid surface, above which lies the atmosphere, a gaseous layer made up principally of nitrogen (78%) and oxygen (21%), which provides us with air to breathe and also protects us from the harmful effects of ultraviolet and other short-wave radiation from the Sun. Much of the Earth's surface is covered in water, a commodity vital to the existence of life as we know it, and the oceans, together with the few metres of surface soil, provide the major habitat for the Earth's creatures.

The mean temperature of the Earth is about 12°C, well above the freezing point of water, but in places the temperature can reach 50°C while in others it may drop down below −50°C. Under these conditions an extraordinary variety of species live and coexist. But our environment is fragile—a small change in the level of solar radiation, which would raise or lower the mean temperature, could produce radical changes that might eliminate advanced species. Man himself is altering his environment at a rapid rate with consequences that can only dimly be foreseen.

Compared to the Sun, the Earth is insignificant. The Sun is a star, and as such it produces energy by means of thermonuclear reactions going on in its central core. It is made up primarily of the lightest and commonest elements in the Universe, hydrogen and helium, and it is a hundred times larger in diameter than the Earth; it has a mass some 330,000 times greater than our planet. Its distance from us is about 150 million kilometres; light, which travels at a speed of 300,000 kilometres per second, requires about 8.3 minutes to reach us from the Sun.

At its surface, the Sun has a temperature of about 6000°C, but deep in the central regions the temperature is believed to be as high as 14 million degrees Centigrade*; at the very high temperatures and

---

*Astronomers and physicists usually measure temperature on the Absolute, or Kelvin, temperature scale which begins at **absolute zero,** the lowest possible temperature, which is equivalent to about −273°C. One degree on this scale is the same 'size' as one degree on the Centigrade scale, but is denoted by the term **Kelvin,** or K. For example, 373 K is the same as 100°C, and 14,000,000°C is, near enough, the same as 14,000,000 K.

pressures prevailing there, hydrogen is fused together to form helium, and this process of fusion releases vast amounts of energy. In each nuclear reaction, what essentially happens is that the nuclei of four hydrogen atoms combine to form the nucleus of one helium atom, and in the process a certain amount of mass is destroyed. This 'lost' matter is converted into energy according to a relationship which derives from the Special Theory of Relativity. In this way the Sun is losing mass at a rate of rather more than four million tonnes *per second* in order to sustain its present power output of nearly four hundred million million million million watts (i.e., 4×10^{26} watts). This is a vast outpouring of energy per second and, although not much more than a ten-billionth of this energy falls on the Earth, even that is still ten thousand times greater than the present energy consumption of the human race.

Around the Sun revolve at least nine planets, some of which have attendant satellites (such as the Moon, in Earth's case), which, together with a host of minor bodies—the asteroids, comets, meteoroids, interplanetary gas and dust—make up the Solar System. In order of distance from the Sun, the planets are Mercury, Venus, Earth, Mars, Jupiter, Saturn, Uranus, Neptune and Pluto. Mercury, about two-fifths of our distance from the Sun, revolves around it in a period of 88 days, while distant Pluto, at 40 times the Earth's distance, requires 248 years to complete one circuit of the perimeter of the planetary system. The properties of the planets and their orbits (paths) around the Sun are listed in Table 1.

The first four planets, Mercury to Mars, are called the terrestrial planets because—although they differ in detail—they are basically similar to the Earth in the sense that they are compact dense bodies with solid surfaces. The next four, the giant planets, are quite different, being composed of lighter elements, having low densities and, so far as we can tell, not possessing solid surfaces. Jupiter is the largest planet of all, being more than twice as massive as all the others put together; it is composed largely of hydrogen and helium and is probably liquid throughout most of its globe, although having a gaseous atmosphere thousands of kilometres thick. Even so, Jupiter has less than one-thousandth of the Sun's mass. If we could observe the Sun from the distance of the nearest star, then with sophisticated instruments we *might* be able to detect Jupiter and possibly even Saturn, the second largest planet of the Solar System, but it is most unlikely that we could detect the other planets, and certainly not the Earth. Perhaps this helps to set the Earth in perspective.

Of the various minor members of the Solar System, some mention should be made of asteroids and comets. The asteroids—more appropriately known as the minor planets—are tiny bodies ranging in diameter from about 1000 kilometres to less than one kilometre; most of them pursue orbits that lie between those of Mars and of Jupiter. The largest are more or less spherical bodies, but most of them are irregular rocky

TABLE 1

PLANETARY DATA

Planet	Mean distance from Sun in astronomical units	Mean distance from Sun in millions of kilometres	Orbital period	Axial rotation period	Equatorial radius (km)	Mass (Earth=1)	Mean density (Earth=1)	Surface gravity (Earth=1)	Number of satellites
Mercury	0·39	57·9	87·97d	59d	2430	0·055	1·0	0·36	0
Venus	0·72	108·2	224·7d	243d(R)	6070	0·82	0·94	0·87	0
Earth	1·00	149·6	365·26d	23h 56m	6378	1	1	1	1
Mars	1·52	227·9	687·0d	24h 37m	3395	0·11	0·71	0·38	2
Jupiter	5·20	778	11·86y	9h 50m	71,300	317·8	0·24	2·64	14
Saturn	9·54	1427	29·46y	10h 14m	60,400	95·2	0·13	1·15	10
Uranus	19·2	2870	84·01y	10h 49m(a)	24,000	14·5	0·29	1·15	5
Neptune	30·1	4499	164·8y	15h 48m(b)	23,000	17·2	0·41	1·4	2
Pluto	39·5	5900	247·7y	6·3d	4000 ?(c)	0·2 ?	1 ?	?	–

Notes

In columns involving time measurement, y=year, d=day, h=hour and m=minute.

(R) implies retrograde rotation: Venus rotates in the opposite direction to the other planets.

(a) Recent estimates suggest that the rotation period may be longer than this commonly quoted value, possibly lying in the range 17–22 hours.

(b) Recent estimates suggest the rotation period of Neptune may be about 23 hours.

(c) The radius of Pluto is not well known, estimates ranging between 3000 and 6000km; few of Pluto's physical properties are known with any certainty.

and rocky–metallic fragments. The origin of these bodies is not absolutely certain—they probably represent debris left from the formation
of the Solar System, but some have speculated that they are the fragments of a shattered world. It is estimated that perhaps a hundred
thousand asteroids exist in all. A few of these have orbits which cross
that of the Earth, and it is possible—although highly unlikely—that the
Earth could suffer a collision with one at some future date. Some, too,
pass beyond the orbit of Jupiter, and in 1977 Charles Kowal at the
Mount Palomar Observatory discovered a hitherto unknown large
asteroid-type body (a 'tenth planet') between Saturn and Uranus;
perhaps there are many more of these bodies than had previously been
imagined.

From the point of view of the exploitation of space, the asteroids
could be particularly important. The average iron content of meteorites
(lumps of rock and metal which hit the Earth) is about 27%, and if the
asteroids are of similar composition then they contain an abundance of
useful material which is easily accessible for two reasons—the gravitational field of most asteroids is negligibly small, so that spacecraft need
only rendezvous with them rather than 'landing' and 'taking off' with
great expenditure of fuel; and, because asteroids are so small, all their
minerals are close to the surface. They offer the prospect of the ultimate
in open-cast mining—the whittling away of entire celestial bodies.
From the point of view of interstellar travellers, asteroids—if they are a
common feature of planetary systems—offer the means by which colonists could obtain mineral resources without disturbing the biospheres of
the planets in their target system.

Comets are rather insubstantial objects. A bright comet can be a
spectacular phenomenon, having a bright head from which flows a
luminous tail, but it is generally reckoned that a comet consists essentially of an icy nucleus, possibly only a few kilometres in radius,
surrounded by a cloud of gas and dust. Most comets pursue very
elongated orbits round the Sun and it is only when they approach close
to the Sun that material is vapourized, streaming away to form the
visible tail. It is considered that most comets spend the greatest part of
their existence well beyond the orbit of Pluto and it is speculated that
there is a 'comet cloud' surrounding the Solar System out to one
hundred thousand times the Earth's distance from the Sun. Comets are
natural probes of interstellar space and the study of their composition
may tell us much about the material out of which the Sun and Earth
were formed in the past. Since comets contain the basic materials
essential to the formation of life there has been much discussion recently
as to whether comets falling on the Earth in the early stages of the
Earth's history provided the essential elements for the origin of life
here; perhaps comets themselves are a suitable environment for life!

When we leave the confines of the Solar System we have to deal with
distances so vast that to talk of them in conventional terms (miles,

kilometres or whatever) becomes quite meaningless to most of us. In many ways it is easier to think of these distances in terms of how long it would take for a signal to cross them. The fastest signal we can send is in the form of light or radio waves, both of which are forms of electro-magnetic radiation, which moves through space at the speed of light, about 300,000 kilometres per second. In one year a ray of light will cover 9.5 million million kilometres, and this distance is known as one light year. In these terms, the nearest star (apart from the Sun itself) lies at a distance of 4.3 light years, and a ray of light will take 4.3 years to cover that distance. By way of comparison light requires just over 5 hours to reach us from Pluto, 8.3 minutes to reach us from the Sun, and only 1.3 *seconds* to reach us from our nearest neighbour, the Moon. In fact, astronomers today use a rather larger unit to measure inter-stellar distances: the parsec (pc), which is equivalent to about 3.26 light years.

The nearest star is Proxima Centauri, a dull red star too faint to be seen without telescopic aid. Stars differ from each other enormously in their properties. Their masses range from less than one-tenth of the Sun's value to nearly a hundred times the solar mass; in luminosity (the amount of light which they emit from their surfaces) they range from less than one-hundred-thousandth of to well over one hundred thousand times the Sun's luminosity. At its surface, the Sun has a temperature of nearly 6000 K; some stars are hotter than 30,000 K while others are cooler than 3000 K. The temperature of a star is indicated by its colour. Red stars are cool (a good example is the star Betelgeuse in the north-east corner of the constellation of Orion); orange stars are somewhat hotter; yellow stars like the Sun are hotter again (about 6000 K); white stars such as Sirius are hotter still (about 10,000 K); while the hottest stars of all appear blue. Astronomers classify stars, according to the appearance of their spectra,* into the following principal classes (in order of descending temperature): O, B, A, F, G, K, M, R, N, S. Each individual class is further subdivided into ten parts, numbered 0 to 9. Thus the Sun is a star of spectral type G2, and as such is a very middle-of-the-road star.

The Sun has a radius of 700,000 kilometres. There exist stars hun-dreds or even thousands of times larger; for example (again), Betel-geuse, which—if placed where the Sun is—would contain the orbits of all the planets out to and including Mars. Betelgeuse is a type of star which we call a red giant; there exist stars known as white dwarfs which are only about the size of the Earth. Even smaller are the collapsed stars, neutron stars, which may be less than 10 kilometres in radius, yet still contain as much material as the Sun.

Some stars are constant in their light output, others vary; some are single (like the Sun), others are double (binary—a pair of stars travel-

*The spectrum of a star is the pattern of light which results when the star's light is spread out into its constituent wavelengths by means of a spectroscope.

ling around each other); and still others belong to complex multiple star systems.

Between the stars lies an extremely thin mixture of gas (mostly hydrogen and helium) and dust (small particles of solid material). The gas sometimes shows up as luminous clouds, known as emission nebulae, such as the North America Nebula (Plate 7), while dense dust clouds show up as dark patches against the starry background. The material is very tenuous indeed, the average overall density of interstellar matter being about 10^{-21} kilograms per cubic metre (kg/m³), about one thousand million million millionth of the density of air at ground level. In effect this means that we should expect to find only one hydrogen atom in a cubic centimetre of space, while a cubic kilometre of interstellar space would contain only a few hundred tiny dust grains.

These densities are far less than the density of the best vacuum Man has produced on Earth, but, added up over the vast volume of interstellar space, both the gas and the dust are very significant. The dust acts as a kind of celestial fog, limiting the distance to which we can see in certain directions in space, while the denser gas clouds provide the material out of which new stars are forming even today. As we shall see in Chapter 6, both gas and dust are significant quantities so far as interstellar travel is concerned, particularly if speeds approaching that of light are contemplated.

Stars form out of gas clouds, contracting until they become hot enough and dense enough for fusion reaction to produce energy. This halts the contraction as the star becomes a stable main-sequence star; this is the stage at which the Sun lies in its evolution, and the stage at which each star spends the major part of its life. As the central supply of fuel dwindles, a star swells up for a relatively brief period to become a red giant before finally consuming all its available fuel and ending its days as a dense white dwarf which cools down and fades away to become, ultimately, a dense dark body. More massive stars may end up as neutron stars—or even black holes, the most enigmatic of astrophysical objects.

The lifespan of a star is decided primarily by its mass—the more massive the star, the shorter its lifetime. This may seem strange, since surely a more massive star has more fuel; but the star consumes that fuel more rapidly. For example, a star ten times the Sun's mass will be several thousand times more luminous than our Sun, and so will consume fuel thousands of times faster than our Sun; with the result that it will live for less than 1% of the Sun's estimated lifetime. The Sun is expected to spend some 10 billion years as a stable main-sequence star, whereas the most massive stars of all—the hot O- and B-type stars—will live for only a few million years.

Stars, their planets and the gas and the dust which we have described form part of one giant star-system that we call the Galaxy. About one hundred billion (10^{11}) stars are contained in a thin disc-

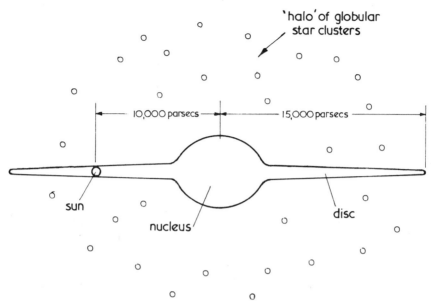

Fig 1 **Our Galaxy in cross-section.** The whole system is about 30,000 parsecs in diameter, but the disc is only about 1000 parsecs thick. Our Sun lies about 10,000 parsecs from the galactic hub, roughly at the centre of the circle shown.

shaped system nearly thirty thousand parsecs in diameter (fig. 1), and the Sun lies about three-fifths of the way from the centre to the outside edge. The galactic centre, or nucleus, contains a dense concentration of stars, but also contains several compact and rather puzzling powerful energy sources whose nature remains something of a mystery. The Sun and Solar System revolve around the centre of the Galaxy in a period of about 220 million years, a period of time sometimes known as a 'cosmic year'. One cosmic year ago, the present continents were still joined together, and the great era of the dinosaurs had yet to dawn. Dare we speculate on the state of the Earth in one cosmic year's time?

Beyond our Galaxy we can see billions of other galaxies, some similar to our own, others quite different in size and structure. One of our nearest neighbours is the Andromeda galaxy, otherwise known as M31 (Plate 8). It lies at a distance of 675,000 parsecs, yet, despite this great distance, it can be seen with the naked eye under ideal conditions (fig. 2). Galaxies have been observed at distances well in excess of 1.5 billion parsecs and there are objects called quasars which can be detected at even greater ranges—over three billion parsecs in some cases. Such is the scale of the Universe which lies within range of our telescopes.

Our observations show that all the galaxies are rushing away from each other (or, at least, all the clusters of galaxies are moving away

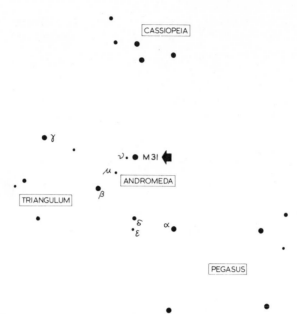

Fig 2 **Locating the Andromeda Galaxy (M31).** The first step is to find the four similarly bright stars which make up the 'square' of Pegasus, a constellation best seen in autumn skies in the northern hemisphere (spring skies in the southern hemisphere). The star at the north-east corner of the square is actually Alpha Andromedae. By looking roughly north-east from this star a distance equal to the length of a side of the 'square', a similar star, Beta (β) Andromedae will be found, and by scanning at right angles to the α–β line past the two faint stars μ and ν, the Galaxy will be located. It is clearly visible in binoculars.

from each other, for galaxies tend to be gathered together in groups). This implies that the whole Universe is expanding, and the best evidence which we have *at present* indicates that the Universe as we know it originated in a tremendous explosion, the Big Bang, which occurred thirteen to twenty billion years ago. All the material which subsequently cooled to form galaxies, stars and planets was hurled outwards from an initial searingly hot fireball, and the fact that the galaxies are seen to be moving apart still is testimony to the violence of that initial explosion. Whether the Universe will continue to expand without limit, or whether it will slow down, stop expanding and then fall together again is but one of the unsolved mysteries of our time. And there is no certainty that the Big-Bang hypothesis is a correct description of the origin of the Universe; it may yet be that some new theory will arise to supplant it.

The Universe is an exciting place, and at times a violent place. There is so much to explore, and so many questions to be answered. Mankind will surely continue to respond to this challenge as it has done in the past.

2 Are we alone?

Of all the questions which can be asked about our Universe, this surely must rank as one of the greatest. Is life unique to the Earth, or do we inhabit a Universe teeming with life of infinite variety? Are we the only intelligent species possessed of the technology which already can allow us to communicate over interstellar distances and which may soon give us the means to travel to the stars, or is intelligence commonplace and interstellar travel and communication already established by civilizations more advanced than ourselves? Will we one day meet our peers among the stars? These questions excite the widest interest and must provide one prime motivation for exploring the Universe.

If we had a clear idea of how life originated on Earth then we might be better placed to answer the question which the title of this chapter poses. The lamentable fact is that we cannot even produce an unequivocal definition of what constitutes life, a definition which would allow us precisely to distinguish between animate and inanimate matter. Indeed, there are some definitions of life which, strictly interpreted, exclude ourselves! In thinking about life elsewhere in the Universe we can only debate sensibly the chances of finding life as we know it, having the same basic structure, composition and metabolism as we find here on Earth. There may exist life of a wholly different kind, having an entirely different basis from anything that we can even imagine, but we can only fantasize about such possibilities. By debating the prevalence of life as we know it, then we are assessing the lower limit of the probability of life elsewhere.

Assuming that we all have some intuitive idea of what constitutes life, let us think about what is required for life as we know it. There are certain basic chemical elements involved, notably carbon, nitrogen, oxygen and hydrogen—and it is significant that these elements are highly abundant in the Universe. It is the ability of the carbon atom to link with others to form large complicated molecules and chains of molecules that allows living species to be built up. Molecules comprising these ingredients are termed organic molecules; one of the most surprising developments in astronomy over the past decade has been the discovery of a wide range of organic molecules existing in interstellar gas clouds. Something like forty different types have been identified, including water, hydrogen cyanide, formaldehyde and various forms of alcohol. In one gas cloud close to the galactic centre there have been discovered sufficient quantities of ethyl alcohol to make something like ten thousand million million million million measures of proof spirit, and those who doubt the value of interstellar travel must surely be swayed by such a possibility!

The discovery of these molecules lends credence to the possibility that life may form far away from planetary surfaces, but, for the purpose of the present discussion, we shall assume that planets are necessary for life. Some kind of energy source is essential, and we may assume that stars provide the basic energy resources, just as the Sun does for us. Water (or an equivalent solvent) appears to be essential. Although there are low forms of life on Earth which can survive prolonged periods in the absence of water, it does appear as if all species require at least intermittent exposure to water—which does, after all, make up about 80% of the content of living cells.

Temperature is important, too. If we assume that water is required in liquid form, then the planet must have such conditions of pressure and temperature as ensure that liquid water is available at least intermittently, and a living creature must be able to maintain its bodily temperature somewhere between the boiling and freezing points of water. An atmosphere is important, if only to shield living material from harmful radiation.

Within the Solar System we can define a region around the Sun—known as the ecosphere—within which the level of solar radiation could sustain a suitable temperature for life; atmospheric conditions may modify this considerably. Within this region lie the planets Venus, Earth and Mars. Mercury, closer in, is too hot, while the more distant planets are probably too cold for life as we know it. Venus, as American fly-bys and Soviet landing missions have shown, is utterly hostile with a temperature of around 750 K at its surface, a dense carbon dioxide atmosphere exerting a pressure at ground level almost a hundred times greater than that which we experience on Earth, and thick unbroken cloud cover, the clouds being made up of droplets of, amongst other noxious chemicals, sulphuric acid; any visiting astronaut who stepped out onto the surface of Venus would find himself simultaneously crushed, incinerated and dissolved. Yet, despite its cauldron-like nature, it is not entirely inconceivable that Venus could support life, possibly in the form of airborne bacteria floating high in the atmosphere.

Despite the remark made in the early part of this century by that notable astronomer—and builder of the Flagstaff Observatory in Arizona—Percival Lowell, 'That Mars is inhabited by beings of some sort or other we may consider as certain as it is uncertain what these beings may be', by the dawn of the Space Age speculation about possible life on Mars was largely confined to bacteria and low forms of plant life. As a result of the fly-by, orbiter and lander missions successfully carried out since 1965 by NASA, we now have a pretty comprehensive view of the red planet. It is a rocky, dusty world having a thin atmosphere largely composed of carbon dioxide and exerting a ground pressure of about 8 millibars, less than 1% of the Earth's atmospheric pressure. Small quantities of nitrogen, water vapour, oxygen and argon are present, but the atmosphere is quite unbreathable to us, and would offer little protection against harmful ultraviolet radiation. There *is* water, locked up in the polar ice caps and in the surface rocks, and although there is no detectable liquid water on the surface there exist a multitude of surface features whose appearance is best explained by assuming that they resulted from the effects of water erosion. There may have been surface water in the past, and perhaps Mars' great system of volcanoes provided the source of atmospheric water in past ages. Although Mars is unpleasant by our standards, it is by no means as hostile as Venus.

In 1976 there landed on Mars two American spacecraft, Viking 1 and Viking 2, whose primary task was to search the Martian soil for any evidence of living material. The results have been inconclusive and perplexing in that some of the experiments have given results which could be interpreted as implying the existence of living matter, while others on the same surface material show no evidence whatever of life. The possibility of the existence of some low form of life on Mars remains for the moment an open question, but there is no doubt that advanced lifeforms are absent.

The rest of the Solar System seems decidedly unpromising, but we should not be too dogmatic about this. For example, giant Jupiter is made up mostly of hydrogen and helium and has large quantities of ammonia and methane in its atmosphere; in fact, the atmosphere of Jupiter is rather like what we believe the Earth's atmosphere to have been before life emerged on Earth. Although the temperature at the top of its cloudy atmosphere is low, about 140 K, a curious thing about Jupiter is that it emits far more heat than it receives from the Sun; this implies that Jupiter must be very hot inside, and at some level below the atmosphere the temperature could be ideally suited to some kind of life. It has been speculated that a lifeform may exist there using ammonia as a solvent instead of water. I am not optimistic about the chances, but the idea of flat fish-like creatures swimming in a sea of ammonia does have a certain appeal.

In the light of what we know about the Solar System it seems certain that if we are seeking advanced forms of life then we must look to the planets of other stars. Current theories of the formation of planets suggest that they form as a result of the same process by which stars themselves are formed; i.e., that they form out of the shell of gas that surrounds a newly forming star as it is contracting out of an interstellar gas cloud. It seems likely that a large proportion of all stars have planets, and this supposition is bolstered by observations which indicate the presence of massive planets orbiting some of the nearest stars. There should be no shortage of planets in the Galaxy.

To make an estimate of the number of life-bearing planets in our Galaxy we need to consider a number of factors, and a formula which attempts to itemize and link together the main factors has been devised by the radio astronomer Frank Drake who, in 1960, instigated the first systematic search for signals from other intelligences (Project Ozma). This equation forms a most useful basis for debate, and is:

$$N = R_* f_p n_e f_l f_i f_c L,$$

where

N = the number of advanced technological civilizations which exist *at the moment* and which have the capacity for interstellar communication,

R_* is the mean annual rate of star formation over the lifetime of the Galaxy,

f_p is the fraction of stars which have planetary systems,

n_e is the average number of planets per system having conditions suitable for life,

f_l is the fraction of suitable planets on which life actually develops,

f_i is the fraction of life-bearing planets on which intelligence develops,

f_c is the fraction of these planets on which the technology for interstellar communication (e.g., radio astronomy) arises, and

L = the average lifetime of such a technological civilization.

The accuracy with which we can estimate these factors declines rapidly as we go down the list; R_* is known to be about ten stars per annum while L is a matter of total conjecture. Let us try to estimate these factors, to see what happens. We shall take the following estimates of the various factors (the reader should try making his own estimates):

$R_* = 10, f_p = 0.1$ (this assumes one star in ten has planets, and is probably an underestimate), $n_e = 0.1$ (if there are ten planets per system, this supposes one in a hundred planets is suitable), $f_l = 1$ (there is a strong body of opinion to the effect that if conditions are suitable, life will develop), $f_i = 0.1$ (we do not know how intelligence developed on Earth but it was presumably by a natural process likely to occur elsewhere), and $f_c = 0.1$ (for no good reason!).

We find that $N = 10 \times 0.1 \times 0.1 \times 1 \times 0.1 \times 0.1 \times L$; i.e., $N = 10^{-3} L$ (i.e., $L/1000$). This implies that one new technological civilization arises in our Galaxy every thousand years, on average.

To estimate N we need to know L, and here we have a problem. We have only one example to base our guess upon, ourselves. We have been advanced, in the sense of possessing radio communication, for less than 100 years, and it is possible that we may annihilate ourselves before much longer. If this is typical of technological civilizations, then we may take L to equal 100 years, and we conclude that we are almost certainly the only advanced technological society in the Galaxy at present. If other species can overcome the problems of resources and social organization then they may survive indefinitely. If, for the sake of argument, we take L to equal 100 million years (10^8 years), then N turns out to be 10^5; i.e., there may be 100 thousand advanced technological communities in the Galaxy at present. The figures are, of course, guesswork, but without making any unduly optimistic estimates (apart from the sheer speculation concerning L) it is not hard to arrive at the conclusion that there may well be a substantial number of advanced species in our Galaxy.

Intelligent life aside, the chances of life of some kind existing must be overwhelmingly high. Using the estimates above (of R_*, f_p, n_e and f_l), we find that the number of life-bearing planets in the Galaxy may be about 0.1 times the number of years for which life of the most basic kind can endure. In the case of the Earth, life has existed for over three billion years and, barring some major disaster, *some* kind of life should persist for a further five billion years, until the Sun becomes a red giant. Being conservative again, let us say that basic lifeforms exist for one billion years on the average planet. Then, the number of life-bearing planets in our Galaxy becomes $0.1 \times 10^9 = 10^8$, or 100 million. This, I am convinced, is probably an underestimate, and, by the time we take into account the billions of other galaxies, the number of possible abodes of life in the observable Universe becomes very high indeed.

In exploring the Universe it seems almost certain that we shall find life of some kind; and, even if we don't, that of itself would be singularly interesting. But there must be a significant chance that in our explorations we shall encounter, in time, other advanced beings, or the remains of past civilizations. Are we alone? I doubt it, but it is a question which requires for its solution the further outward-looking exploration of mankind.

3 The Why of it All

Let me try to summarize the reasons why I feel it to be almost inevitable that if mankind survives the next century it will reach out towards the stars. We have already seen that there is a certain historical inevitability about it, for Man has exhaustively explored and exploited the planet Earth and is already moving his sphere of activities out into near-Earth space. There seems no reason, barring disaster, for this process to come to an end. The pressures that are only too obviously arising from Man's confinement on a world of limited resources will force upon us a number of options, and with every day that passes the likelihood of our being able to *choose* our future diminishes. The major alternatives which lie ahead are conditioned by whether or not we come to grips promptly with the problem of rising world population. If we do, then we may choose either the inward-looking option of accepting the limitations of our terrestrial environment and operating a carefully-regulated and restricted way of life on a global basis, or the outward-looking option of expanding Man's activities into space. If we do not come to grips with the population problem then there can be no doubt that a major catastrophe lies in store for the human race (catastrophe may lie ahead in any case, but the establishment of zero population growth would greatly aid its avoidance).

Expanding human activity into space on a large scale would give us access to a constant, cheap and reliable source of power in the shape of the Sun, to new mineral resources on the Moon and among the asteroids, and to an environment highly suitable for specialized technologies (high vacuum and zero gravity, for example, are both conditions difficult or impossible to attain on Earth). Economic growth and technological advancement could continue without further damaging the fragile balance of the Earth's environment. Solar power beamed to the Earth could make a substantial contribution to the energy crisis with far less environmental impact than other major solutions currently being debated. With the appropriate attitude of mind the growing wealth of humanity could be used for the wider benefit of humanity, and it is difficult to see how a global population of 6 billion (which will be the figure in the year 2000—even assuming an imminent end to population growth) could be supported at even a modest standard of living without technological developments and some industrial activity.

Given the large-scale humanization of the Solar System, wide-ranging exploration and the establishment of colonies in space and on the planets, Man will become at home in his new space environment, and must surely wish to explore it further. Given this situation he will have the economic resources and, given some development of existing technology, he will have the technical means to undertake interstellar exploration. With his innate urge to explore and his deep-seated curiosity about his surroundings, I cannot believe that Man would be content to remain within the Solar System.

Moving into the Solar System would remove the restrictions imposed by mankind's being confined to a planet, but even the Solar System is restricted in the volume it offers for exploration. Only by launching ourselves forth into interstellar space can we free ourselves from all known physical boundaries; space, so far as we know, offers us an infinite frontier. We may in time find new limits to our activities, but these limits, if they exist, lie beyond our present comprehension and well beyond the range attainable for the foreseeable future. Such limits can safely be left to be handled by a wiser and more diversified human species of the future.

By expanding the range of human habitats we should be taking positive steps to ensure the survival of the human species. At the moment mankind is restricted to one habitat, the Earth, and is vulnerable to complete extinction as a result of some global disaster—man-made or otherwise. By establishing human communities throughout the Solar System, mankind is secure from a single catastrophe of that nature, but even the Solar System itself may not offer sufficient long-term security. For example, from time to time the Sun passes through dense interstellar dust clouds with resultant changes in its output of energy; this could have disastrous consequences for life. By extending human habitats into interstellar space, the long-term future of the species is better assured.

The establishment of human colonies in diverse and widespread environments would lead to physical, mental and cultural evolution in multitudinous ways. Variety is a vital element in the survival of an ecosystem, and there is every reason to suppose that variety in human colonies is equally essential to the survival of the human species, and to whatever may evolve from that *Homo sapiens* who climbed out of his terrestrial cradle.

In expanding our sphere of influence among the stars we may encounter other intelligences and other technologies. The effects of such contact are discussed in Chapter 11, but at the moment it seems most likely that, if we do make such an encounter within the next thousand years, it is more likely to be with a civilization far in advance of our own, rather than with one which is less advanced. The effects of such an encounter may be for good or ill, but any ill consequences could scarcely be avoided by hiding away on Earth and hoping that some other exploring alien would not find us. If, perchance, we are the most advanced species in the Galaxy, it is in our interests to expand our sphere of influence and to monitor developments elsewhere. If we are the only intelligent species, surely we have a *duty* to explore this Universe.

It may be argued, with some justification, that we could carry out an extensive exploration of the Universe without resorting to manned interstellar flight, but Man as he is today requires a physical challenge as well as an intellectual one. This attitude may change with time, but such a change cannot be foreseen on the basis of our present knowledge of ourselves. Barring such a fundamental change in human nature, I am convinced that, given the means, some members of the human community would wish to travel along the road to the stars. Physical challenge provides an essential safety valve to mankind.

In the end it will be the urge to explore and to answer those questions which, having been answered, lead on to further, deeper questions, that will beckon mankind on to the 'final frontier'. For the moment, space is our ultimate frontier; in the future there may be others, more subtle perhaps. But it is to the space frontier that this book is addressed, and to ways in which we may respond to its irresistible challenge.

PART II

Ways and Means

4 Home Territory:
(i) the story to date

When Neil Armstrong and Edwin Aldrin gingerly stepped down from the lunar module *Eagle* onto the Sea of Tranquillity to become the first men to set foot on another world, the rest of humanity waited tensely to see the outcome. I well remember sitting up through the night of 20 July, 1969, into the following morning, for it was at 03 56 (British Summer Time) that Neil Armstrong placed his boot in the lunar dust and made that slightly indistinct and oft-misquoted remark which will go down forever in human history: 'That's one small step for [a] man, one giant leap for mankind.' The television picture quality was atrocious by comparison with that of later missions, but those of us watching sensed nevertheless that this truly was a crucial moment in human evolution, the moment when Man ceased to be limited to one planet, when he left his birthplace and took the first faltering step along the path to the stars.

The Apollo landing was the fulfilment of a desire as old, perhaps, as the longing to be able to fly like the birds, and it marked the culmination of centuries of speculation, reason and technological development. It mattered little that the Apollo Project itself owed its origin to a political decision to demonstrate to the rest of the world the technological prowess of the United States of America. The programme was inaugurated on 25 May, 1961, by President John F. Kennedy with these words:

> I believe this nation should commit itself to achieving the goal, before this decade is out, of landing a man on the Moon and returning him safely to the Earth. No single space project in this period will be so impressive to mankind, or more important for the long-range exploration of space, and none will be more difficult or expensive to accomplish.

Eight years later that goal was achieved, but changing world circumstances rendered the political capital less valuable than was anticipated. The fact which remains, and which in the long term will prove to be the important point, is that men did travel to another world, setting in train a process which—barring disasters—will lead us on to explore the stars.

That the idea of space travel is not new is evinced by the oft-quoted *True History* written by the Greek satirist Lucian in the second century AD. According to this delightful tale, the heroes of the story were sailing through the pillars of Hercules (the Straits of Gibraltar, in modern parlance) when they were picked up by a waterspout and deposited—to their considerable surprise—on the surface of the Moon.

The seventeenth century, a period of immense scientific and cultural activity, saw the appearance of a number of books with 'space-travel' themes. The great astronomer Johannes Kepler, who finally dispelled the dogma of uniform circular motion in the heavens by showing that the planets move around the Sun in elliptical orbits, was also interested in mystical ideas, and his story, the *Somnium* (the 'Dream'), which was published in 1634, four years after his death, had a decidedly mystical flavour. The hero of the story, Duracotus, was pulled by demons along the track made by the Earth's shadow during an eclipse of the Moon. Once he reached the point where the Moon's attraction balanced that of the Earth, he was allowed to fall towards the Moon. In essence, the story introduced an idea with which we are now familiar, the concept of a 'neutral point' where the gravitational attractions of Earth and Moon are equal and opposite. The effects of being above the air are also discussed.

This altogether remarkable tale was followed in 1638 by Bishop Godwin's *Man in the Moone* (in which the hero was towed to the Moon by a flock of swans) and two years later by John Wilkins' work suggesting that the Moon might be inhabited and that in the future the means might be developed to travel there and colonize it. In the light of recent events he would seem to have been remarkably prescient. Cyrano de Bergerac's *Voyages to the Moon and Sun*, published in 1656, contained some fascinating speculations, including an (unsuccessful) attempt at space travel in which the 'astronaut' strapped bottles of dew round his waist in the hope that the Sun would suck him upwards—after all, is it not an established fact that dew vanishes after the Sun rises? Other suggestions in the book included a form of rocket propulsion in which solar heating caused hot gas to escape from a chamber.

In the eighteenth century came the publication of Jonathan Swift's *Gulliver's Travels*, a work of lasting popularity which included an account of how the astronomers of the flying island of Laputa had discovered two tiny moons circling the planet Mars, a result remarkable for the fact that it preceded the actual discovery of these satellites by about one hundred and fifty years. Later in the same century, in 1752, the great French writer Voltaire produced the story of Micromegas, a giant from the planetary system of Sirius.

Undoubtedly the most remarkable story to emerge from the literature of the nineteenth century was Jules Verne's *From the Earth to the Moon* (1856), which was based on some sound scientific principles and which contained some remarkably accurate speculations. Verne's spaceship and its crew of three was fired from the *Columbiad*, a giant cannon set up in Florida close to Cape Canaveral. The capsule was fired at a speed of 11 kilometres per second, this being the Earth's escape velocity, the minimum velocity at which a projectile must be fired in order that it will leave the Earth and not fall back to the ground. Admittedly, any astronauts who were fired from a cannon at this sort of speed would be crushed to a fine paste on the floor of the spacecraft by the sudden acceleration, but we must allow some artistic licence. After approaching the Moon—but not landing—the projectile returned to Earth, splashing down in the Atlantic Ocean, where the travellers were picked up by an American warship. The parallel with the Apollo 8 mission which occurred 112 years later are remarkable—Verne got the launch site, velocity and landing technique absolutely right.

In the twentieth century stories involving spaceflight have flourished, and as fictional ideas have been outstripped by reality, so the writers have extended their territory far beyond the Solar System to fantastic new worlds, galaxies and universes. The fiction writers of the past have had some remarkable predictive successes (although we tend to forget the multitude of predictions which have not been fulfilled), and I have little doubt that today's writers will do likewise.

By the late nineteenth century it was becoming appreciated that some form of rocket propulsion was necessary for interplanetary flight for, as we shall see later, the rocket (in some form or another) is the only known type of propulsion which can operate in the airless void of space. The screw of a ship operates on the water, pushing the water back and the ship forwards; the airscrew operates in a similar fashion for an aircraft: even the jet engine has to take in air to operate. The rocket needs nothing on which to act; it is self-contained and functions most efficiently in a vacuum.

The first crude rockets were (probably) built by the Chinese over a thousand years ago, and used an explosive mixture like gunpowder; the first records of their being used as weapons of war date back to the seiges of Kiai-fung-fu and Ho-yang in 1232. There are occasional records of subsequent military application, but, in Europe at least, their use tended to be restricted to firework activity until in about 1802 Colonel William Congreve designed a rocket to be used for life-saving purposes (carrying a line to stricken ships). Congreve's improved rockets were used for bombardments carried out during the Napoleonic Wars. Such rockets were solid-fuelled devices, burning an explosive mixture in powder form and, partly because of the difficulty of controlling the rate of burning, tended to be pretty inaccurate and unreliable devices. The advance which led to the rocket's becoming the means of interplanetary travel was the development of the liquid-fuelled rocket, in which fuel and oxidant are fed into a combustion chamber at a controlled rate and react to produce the hot gas which provides the propulsive force.

The true pioneer of this type of rocket was undoubtedly the Russian schoolmaster Konstantin Eduardovitch Tsiolkovskii (1857–1935). In his numerous works he laid down practically all the theoretical foundations for rocket propulsion, pointing out the advantages of the controllable liquid-fuelled rocket, and suggested liquid hydrogen as the ideal fuel. He discussed the merits of building rockets in stages (a concept we shall discuss later in this chapter) and examined the problems of sustaining Man in the hostile environment of space; he further suggested the possibility of man-made colonies in space. It is significant that the Soviet Union should have launched the first artificial satellite in 1957, the centenary of Tsiolkovskii's birth, and it is most appropriate that the most conspicuous crater on the far side of the Moon should have been named in his honour.

1
Venus, photographed in February 1974 by the US space probe Mariner 10. The photograph was taken in ultraviolet light and clearly shows the circulating cloud pattern. (*NASA.*)

2
The four largest Martian volcanoes are visible in this photograph taken by the US Viking 1 craft as it approached Mars in June 1976; the range is 575,000km. The uppermost of the four is Olympus Mons, which has a base 600km across and whose height at the highest point is nearly 25km above the level of the surrounding plain. The great rift Valles Marineris, stretching some 4000km, can be seen faintly in the lower half of the disk between the terminator and the bright limb. The bright area at the bottom is Argyre, a large impact basin. The south pole is in darkness at the lower left of this photograph. (*NASA.*)

overleaf
The Daedalus spacecraft departs from the vicinity of Jupiter as it commences its fifty-year voyage to Barnard's Star. The outcome of several years' work by Daedalus Study Group of the British Interplanetary Society, this starship design is probably the most detailed to date. The spacecraft is propelled by a nuclear pulse rocket using deuterium and helium-3 as propellant. 30,000 tonnes of helium-3 are required and, since helium-3 is very rare in nature, it is necessary to 'mine' it from the helium-rich atmosphere of the giant planet. The space station shown has acted as a base for the construction crew.

After an acceleration period lasting nearly four years, the probe—which is a two-stage vehicle—will attain a cruise velocity of between 12% and 13% of the speed of light. Its payload of 500 tonnes of scientific equipment includes a number of planetary probes that will be despatched prior to the craft's arrival at the target system, nearly two parsecs distant.

1▲   2▼

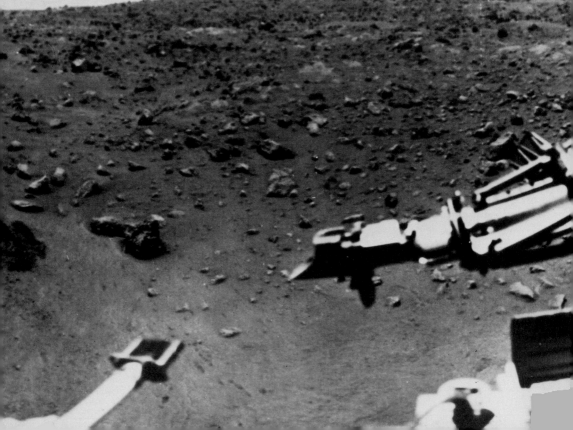

Independently, in the United States, Professor Robert H. Goddard (1882–1945) did much theoretical work on the mechanics of the liquid-fuelled rocket. His classic paper on rocketry appeared in 1919 under the sober scientific title of 'A Method of Reaching Extreme Altitudes', and was instrumental in his receiving some funding from the Smithsonian Institution to carry out practical experiments. The paper also included a brief reference to the possibility that the rocket might be developed to the stage where it could reach the Moon, and it was this suggestion that made headline news in the *New York Times* at the time.

On 16 March, 1926, he launched the first successful liquid-fuelled rocket which, using a combination of kerosene and liquid oxygen (essentially the same propellant which launched John Glenn, the first American to go into orbit, 36 years later), reached a speed of about 100kph and soared to a height of about ten metres. Despite its admittedly modest performance, Goddard's rocket was to astronautics what the Wright brothers' flight of 1903 had been to aeronautics. Nine years after his first attempt he was able to fire rockets to a height of two kilometres.

Sparked off by such writers as Hermann Oberth and Willy Ley, interest in rocketry developed in Germany through the nineteen twenties and 'thirties and led to the emergence of the rocket as a serious military weapon. This work culminated in the V2; nearly 15 metres in length and weighing some 13 tonnes at lift-off, the V2 was capable of carrying a tonne of high explosive over a range of some 300 kilometres, and of reaching a maximum velocity of over 6000kph. At the end of World War II many of the V2 team, including the outstanding rocket

3
A photograph from the US Viking 1 craft, 31,000km above the surface of the planet Mars. The photograph shows the Valles Marineris, nicknamed the 'Grand Canyon of Mars', which lies just a few degrees south of the Martian equator and runs parallel to it for some 4000km: in this picture, the equator cuts across the top left corner, and the north direction lies from lower right to upper left. (*NASA.*)

4
A summer day on the planet Mars, as seen by Viking 1 in August 1976. Towards the left is a small dune of fine-grained material scarred by trenches dug by Viking's surface sampler. The sampler scoop is visible in its 'parked' position. The bright reddish-orange surface of Mars is strewn with a variety of angular rocks of several types. The lighter-coloured patches in the middle and far distance are outcropping areas of the bedrock that underlies the landing site. The cracked and pitted surface at the bottom of the picture was caused by Viking 1's rocket exhaust blast during touchdown. (*NASA.*)

pioneer Wernher von Braun (1912–1977), together with a number of intact rockets, were transported to the United States. Likewise many of the technicians from the launch site (Peenemunde) were captured by Soviet forces, and there is little doubt that the rocket vehicles which launched the early Soviet and American satellites owed much to their predecessor, the V2. The V2 itself was the first rocket vehicle effectively to leave the atmosphere, for in vertical flight it was capable of reaching altitudes well in excess of 150 kilometres.

The subsequent development of the rocket, and of the 'space race', is a vast and complicated subject which has been documented elsewhere. The pace has been spectacular: from Goddard's first liquid-fuelled rocket to the first satellite took a mere 31 years, Sputnik 1 being launched on 4 October, 1957. In that time the speed attained by rocket vehicles increased from below 100kph to above 28,000kph. In January 1959, the Russian probe Luna 1 flew past the Moon, and a couple of years later Yuri Gagarin became the first man in space; his epic flight on 12 April, 1961, in the spaceship Vostok 1, lasted 108 minutes, took him once round the world, and elevated him to a maximum altitude of 327 kilometres. The following month came President Kennedy's commitment to the programme which led to the successful landing of Apollo 11 on the Moon only 43 years after Goddard's first spindly rocket took its own 'small step'.

How the Rocket Works

The rocket works on the principle of reaction—by throwing material in one direction the rocket is caused to move off in the opposite direction. This is an example of Newton's Third Law of Motion, 'to every action there is an equal and opposite reaction'.

Imagine for a moment that you are standing on a sledge on a perfectly smooth sheet of ice, and suppose that you have on board the sledge a stock of blocks of ice—a readily obtained 'fuel' in the circumstances. Let us further assume that the sledge plus crew (you) have a combined mass of 100 kilograms and that you have initially 172 blocks of ice each having a mass of 1kg (the reason for this figure will become apparent as we go along). The total mass, comprising the vehicle, the payload (you) and the 'fuel', is 272kg (fig. 3).

Assuming that you are endowed with incredible strength, fitness and reserves of energy, you find you can throw one of these blocks at a speed of 100 metres per second. The principle of reaction tells us that, if the iceblock shoots off in one direction, the sledge must move off in the opposite direction. Of course the sledge plus load weighs much more than the iceblock (271kg compared to 1kg) and the speed it attains will be less than that of the iceblock by the same proportion; i.e., $100/271$m/s (about 37 *centimetres* per second). If the ice is perfectly smooth, so that there is no friction, and if we ignore air resistance, the sledge will continue to move at that speed indefinitely in accordance

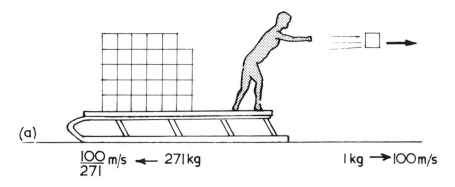

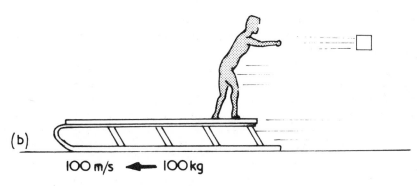

Fig 3 **The principle of reaction.** In (a) the sledge carries a load of 172 blocks of ice, each weighing 1kg; as the man throws one block at a speed of 100 metres per second, so the sledge—which is assumed to be on a friction-free surface—picks up a speed of about 37 *centimetres* per second. In (b), as the final block is thrown, the sledge finally attains a speed of 100 metres per second.

with Newton's First Law: 'Every body continues in its state of rest or of uniform motion in a straight line unless acted upon by an external force.'

If you now throw another block with the same velocity as the first, the sledge will increase its velocity by 100/270m/s; the increase in velocity is slightly greater this time as the sledge plus load weighs slightly less than on the first occasion (270kg compared to 271kg). As each block is thrown off a further impulse is given to the sledge and the speed builds up until, when the final block is thrown, the last velocity increment is 100/100; i.e., 1 metre per second (the weight of the sledge plus load by then being reduced to 100kg). Adding up all the increments we find that the final velocity of the sledge is 100 metres per second, exactly equal to the speed at which each of the 1kg blocks was ejected.

This illustration, although artificial, tells us a good deal about the principle of the rocket which is, after all, simply ejecting particles of gas

at high speed. One point is that the rocket will continue to accelerate
for as long as it is under powered flight (in the analogy, this is the
period during which we can continue to hurl blocks off the sledge);
when the motor is switched off, the vehicle coasts along at a steady rate
unless acted on by some force (e.g., gravity). As the mass of the sledge
plus load decreases, the effect of each constant impulse increases (the
first block accelerated the sledge by 37 centimetres per second, while
the final block increased its velocity by 100 centimetres per second);
thus, if a rocket is subject to a constant force, the rate of acceleration
increases as the fuel supply diminishes. The corollary of this is that
much of the energy released by the fuel* is wasted in accelerating the
rest of the fuel as well as the spacecraft itself. This is a fundamental
weakness of the rocket as a means of propulsion—it must carry its fuel
with it, and most of the fuel is expended in this process. If a rocket
could pick up fuel as it went along, the problem would be relieved (see
page 132).

The speed at which the exhaust gases are expelled from the nozzle of
the rocket motor is known as the *exhaust velocity* and is represented in
our analogy by the speed of the ice blocks (100m/s); and the ratio of
the mass of the rocket at launching (i.e., the combined mass of propellant, rocket structure and payload) to the final mass (when empty of
fuel) is known as the *mass ratio*. Both of these factors are vital to the
performance of a rocket, for the final velocity it attains depends on
both; and increase in either or both results in an increase in the velocity
reached when all fuel has been expended.[†]

If the mass ratio is precisely 2.72 then an ideal rocket will attain a
final velocity precisely equal to its exhaust velocity. This is what
happened to us when our initial mass was 272kg, our final mass 100kg
and the mass ratio 272/100 (=2.72). This basic relationship applies to
all kinds of rocket, from fireworks to Saturn 5s, and to the advanced
systems which may propel us to the stars.

There is another important factor—thrust. The thrust is the accelerating force developed by the rocket and is equal to the product of the
amount of mass expelled from the rocket in unit time (the material
expelled being called the 'reaction mass') and the exhaust velocity.
Thus a rocket might have a very high exhaust velocity but still generate
only a small amount of thrust due to its expelling only small amounts

---

*Strictly speaking we should use the term 'propellant', which implies a mixture of 'fuel' (e.g.
kerosene) and 'oxidant' (e.g., liquid oxygen) which burn together in the combustion chamber
of the rocket motor to release energy.

†For those who are interested in making calculations, if we represent the final velocity by v_f,
the exhaust velocity by c, initial mass by m_i and final mass by m_f, the mass ratio is m_i/m_f and

$$v_f = c \log_e \frac{m_i}{m_f},$$

where e is the natural base of logarithms ($e = 2.72$, approximately). If we require $v_f = c$, then
$1 = \log_e m_i/m_f$, and therefore $m_i/m_f = e = 2.72$.

of mass per unit time. An extreme example of this is the everyday electric torch, which is expelling photons ('particles' of light) in a narrow beam with the highest possible velocity, the speed of light; the thrust exerted on the torch, however, as you will notice if you hold one, is immeasurably small.

In order that a rocket takes off from the ground its thrust must be greater than the weight of the rocket. If the thrust were exactly equal to the rocket's weight, the rocket would sit there balanced over the launch pad until sufficient fuel had been expended to reduce the weight below the value of thrust; needless to say, this fuel would have been wasted. In a more extreme case, if the thrust were less than the *empty* weight of the rocket it would never take off at all but sit on the launch pad until it had burned all its fuel. This sort of thing has happened before now.

Out in space, thrust is less crucial. Admittedly, the thrust generated determines the rate at which the rocket accelerates, but it is the exhaust velocity and mass ratio which fix the final velocity attained. The time taken to build up this velocity may not be too important.

The Step Rocket

Now we come to a basic technical problem. Existing rockets derive their thrust from the chemical reaction between fuel and oxidant in the rocket motor, which releases energy and expels a stream of hot exhaust gases. The best modern rockets of this type can achieve exhaust velocities of the order of 3 or 4 kilometres per second, and a mass ratio of 2.72 is required if the rocket is to attain a final velocity of that magnitude. The minimum speed at which a rocket must be propelled if it is to escape from the Earth (the escape velocity) is just over 11km/s (40,000kph), and the best exhaust velocities are only one third to a quarter of that value. To travel at twice its exhaust velocity a rocket needs a mass ratio of $2.72 \times 2.72 = 7.4$, while to reach four times its exhaust velocity requires a mass ratio of $7.4 \times 7.4 = 55$. To attain escape velocity it would seem that the entire mass of rocket structure, motor, fuel tanks, guidance and control systems and payload could make up only a few percent (less than 2% for a mass ratio of 55) of the total, the remainder comprising fuel. To achieve this would require some uncanny engineering!

The solution adopted is the step rocket, basically a series of rockets stacked one on top of the other. As the first rocket (the first stage) becomes exhausted of fuel, it drops away, carrying with it its unwanted mass, and the second-stage motor ignites to accelerate the lighter remaining part of the vehicle. When it is exhausted it, too, drops away, and the third stage takes over to produce further acceleration. The benefits of this approach can be made clear by a (somewhat extreme) example (fig. 4). Consider a two-stage rocket whose second stage consists of 1 tonne of useful payload, 9 tonnes of rocket (empty mass) and

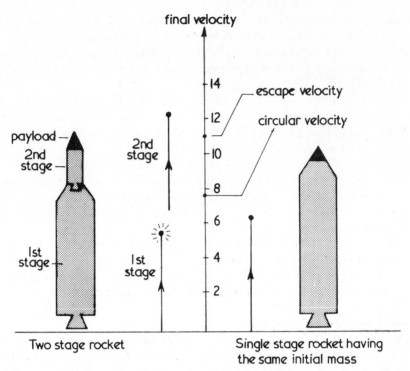

Fig 4 Comparison of two-stage and single-stage rockets. Both rockets have the same initial mass, but the rocket on the left is divided into two stages, each having a mass ratio of 2·72. The two-stage rocket attains a final velocity of 12km/s while the single-stage rocket attains only 6·3km/s.

17 tonnes of propellant. The total mass of that stage is 27 tonnes, and the mass ratio 2.7 (near enough to 2.72). If the first stage is built in proportion, its payload will be the second stage (mass 27 tonnes), its own rocket mass will be $9 \times 27 = 243$ tonnes, and the mass of fuel 460 tonnes in order that it too has a mass ratio of 2.7. The total mass of the two-stage rocket is therefore $460 + 243 + 27 = 730$ tonnes. If the rocket motors had the highly desirable exhaust velocity of 6km/s, the first stage would propel the vehicle up to that speed and drop away, after which the second stage would add a further 6km/s, bringing the final velocity of the 1-tonne payload up to 12km/s, in excess of escape velocity and quite sufficient to reach Mars or Venus.

A single-stage rocket utilizing the same amount of fuel (477 tonnes) and having the same initial mass (730 tonnes) would have a mass ratio of just under 2.9 and would attain a final velocity of only just over 6.3km/s. This is far below escape velocity, and would result in the payload falling back to Earth with an expensive crunch.

Even with step rockets, we require very large mass ratios and exorbitant expenditure of fuel to attain the speeds necessary for inter-planetary travel. If we compare payload with launch-weight for the

Saturn 1B rocket, for example, we find a launch weight of 600 tonnes is required to place 20 tonnes in Earth orbit; the giant Saturn 5 used on the Apollo missions can take 50 tonnes to lunar orbit, but weighs some 3000 tonnes at liftoff (the ratio of initial mass to payload is 60:1). What is required is some kind of fuel with enormously greater exhaust velocities—but there is little hope of any dramatic improvement in the chemical fuels which are now being used. There already exist rocket motors with much higher exhaust velocities—the nuclear and ion rockets which we shall discuss in Chapter 5—but these do not generate enough thrust to get themselves off the ground, and one would not wish to have nuclear reactors flying through the atmosphere in any case. Out in space such rockets have tremendous potential, but for the basic task of getting into orbit we are stuck with chemical rockets for some time to come.

Speed and Energy for Interplanetary Missions

To move into interplanetary space we need to overcome the gravitational attraction of the Earth, and to do that requires a lot of energy. If you were to throw an object upwards with a certain force it would reach a particular height and then fall back again; if you threw it harder, it would reach a greater height before falling down. If you could project something from the Earth's surface at a minimum speed of 11.2km/s then, neglecting the effects of atmospheric friction, it would continue to move away from the Earth and never return. This minimum velocity is known as the escape velocity, and it is easily calculated for any body in the Universe.*

To understand the notion of 'escape' we need to think about the concept of energy (for we cannot truly escape from the Earth's gravitational influence—which spreads out, weakening as the square of distance, to an infinite distance, where, of course, it has negligible effect). A body may possess kinetic energy by virtue of its motion—as any person who catches a fast-moving ball well knows—and it may possess potential energy by virtue of its position—a diver poised on the high board possesses potential energy because of his height above the pool, but may convert this to kinetic energy by the simple expedient of jumping off the board; accelerated by the Earth's gravitational attraction he builds up speed as he falls, and the kinetic energy which he has

---

*Denoting escape velocity by v_e, the mass of the body from which you wish to escape by M, your distance from the centre of the body by R (i.e., if you are on the surface of a planet, $R =$ the radius of the planet), then

$$v_e = \sqrt{\frac{2GM}{R}};$$

where G is the Gravitational Constant which, in Newton's theory of gravity, determines the value of the gravitational force acting between two bodies of known mass and separation. From the formula it can be seen that the value of escape velocity diminishes with increasing distance.

when he hits the water with a splash is just equal to his initial potential energy.

The acceleration experienced by a body falling under the influence of gravity is known—with a far from rare originality—as the *acceleration due to gravity*, and is usually denoted by *g*. Near the surface of the Earth, *g* has a value of about 9.8 metres per second per second (m/s²), which implies that in the first second of a fall a body will acquire a velocity of 9.8m/s, in the next second it will acquire a further 9.8m/s, making a net velocity of 19.6m/s, and so on. Further from the Earth, where the gravitational attraction is weaker, the value of *g* is smaller. If we imagine a body placed at an infinite distance from the Earth and then allowed to fall, it would begin to accelerate very slowly at first, but the acceleration would build up rapidly as it approached close to the Earth. The body would hit the Earth with a kinetic energy exactly equal to the potential energy which it possessed at an infinite distance away. Conversely, if we wish to project a body to an infinite distance, we must give it just this value of kinetic energy. Knowing the relationship between kinetic energy and velocity,* we can easily find the velocity attained by the falling body, which is the escape velocity.

It turns out that the potential energy of a body at an infinite distance from the Earth is just the same as the potential energy which a body at a height above ground equal to the radius of the Earth would have *if* the acceleration due to gravity remained constant with height, rather than diminishing as it does in reality. The escape velocity of the Earth is just the velocity which would be attained under these circumstances by a body released from that height (about 6400km), and it is equivalent in turn to the amount of work done in raising the body to that height. In effect, escaping from the Earth is equivalent to climbing a mountain (under constant *g*) 6400 kilometres high!

This line of reasoning allows us to build a nice analogy of the gravitational fields of Sun and planets. We can think of space away from the planets as being a 'flat' tabletop, and each planet as being at the bottom of a well in the tabletop, the depth of the well corresponding to the strength of the planet's gravitational field or, rather, the amount of effort needed to get away from that planet (fig. 5). Thus the well associated with the Earth is much deeper than the well associated with the Moon, and the escape velocity of the Earth is much greater than the escape velocity of the Moon. By keeping well clear of these wells, it is easy to move things around, but, inside wells, a lot of energy is required to shift things away from the bodies concerned.

A satellite in orbit around the Earth does not need to reach escape velocity; if it is in close Earth orbit (just above the atmosphere) then to

---

*Kinetic energy (ke) = $\frac{1}{2} \times$ mass (*m*) $\times$ velocity squared (v^2). A body of mass 1kg dropped to the Earth from an infinite height would acquire a kinetic energy of about 121,000,000 Joules (the Joule being a unit of energy). This corresponds to a velocity of 11km/s; i.e. to the Earth's escape velocity.

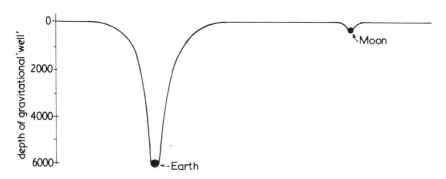

Fig 5 **The gravitational 'wells' of the Earth and the Moon.** Escaping from the Earth is analogous to climbing out of a well 6400km deep (i.e., equal to the radius of the Earth) under a constant gravitational force equal to the value at ground level. To escape from the Moon one requires to climb out of a well which is only some 300km deep.

remain in a circular path it must move at *circular velocity*—just under 8 kilometres per second. The value of circular velocity diminishes with distance from the parent body. We can use the analogy of wells here again—the sides of the well being steeply sloping close to the Earth but flattening out with height, eventually merging into the 'table' (fig. 5). A very fast car can maintain its height on a steeply banked circular race track: if its speed drops, it will drop down; if its speed increases, the car will rise and perhaps overshoot the top of the bank. Likewise a satellite must move sufficiently fast to stay up on the slope of the 'well', but higher up—where the slope is gentler—the required velocity is lower.

The analogy is fine, but what is it that *really* keeps a satellite up there and prevents the Moon from falling onto our heads? Well, the satellite is, in fact, 'falling' all the time (fig. 6). According to Newton's First Law (page 47), a satellite moving at 8km/s at a tangent to the Earth's surface, in the absence of the Earth's gravitational attraction, would carry on in that direction at that speed. The Earth's attraction acts at right angles to the motion of the satellite and is directed towards the centre of the Earth. If the satellite is moving too slowly, it falls and hits the Earth at some point, but, if it is moving at just the right velocity (the circular velocity), then the combination of its 'falling' and its 'sideways' motion results in its maintaining a constant altitude above the globe of the Earth.

A small increase in velocity (fig. 7) changes the orbit into an ellipse, the point of closest approach to the Earth being called 'perigee' and the point of greatest distance 'apogee'. A greater increase causes the satellite to move in a more elongated ellipse, and, if the velocity is increased to escape velocity, then the path becomes a parabola, a curve which is open and does not return to its starting point. Any further

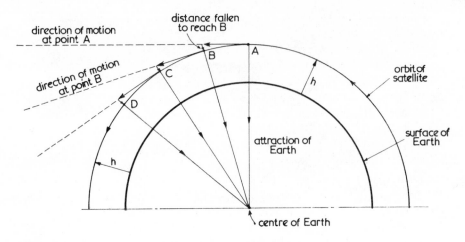

Fig 6 **How a satellite stays in orbit.** At point A the satellite is moving in the direction of the arrow and, if the Earth's gravitational attraction were removed at that instant, the satellite would continue to move at uniform velocity in a straight line as shown. The effect of the Earth's attraction is to cause the satellite to fall towards the Earth and the combination of its motion and this 'falling' is to bring it to point B. The net result of the Earth's attraction and the satellite's transverse ('sideways') motion is to keep it at a constant height above the Earth's surface; i.e., to maintain it in a circular orbit, *provided that* the transverse velocity has precisely the correct value.

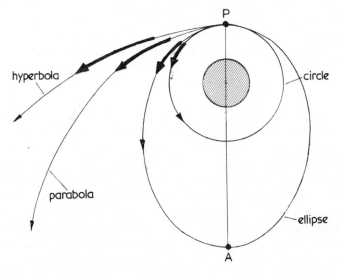

Fig 7 **Conic orbits.** If a spacecraft has a velocity precisely equal to circular velocity, then it will follow a circular orbit. If the velocity is increased somewhat at the point P then the satellite will pursue an elliptical orbit where P is *perigee*, and A is *apogee*. Increasing the velocity to escape velocity will result in the spacecraft taking a parabolic path, while still further increase results in a hyperbolic trajectory. The length of the arrow on each orbit gives an indication of the relative speeds involved.

increase in velocity leads to a hyperbolic trajectory. Isaac Newton was aware of this, and even drew a diagram to illustrate different kinds of orbits in his *Principia Mathematica Philosophiae Naturalis*, published in 1687. It required two hundred and seventy years to develop the technology to utilize the theory!

In a closed orbit—like an ellipse or a circle—the kinetic energy of the spacecraft is less than its potential energy; in a parabolic trajectory, the two are exactly equal—this implies that, when the spacecraft has staggered out 'to infinity', it will have zero kinetic energy and, therefore, zero velocity. 'Infinity', from the astronautical point of view, can be regarded as a few million kilometres when considering escape from the Earth, and a spacecraft launched from the Earth at precisely escape velocity will be reduced to crawling pace by the time it gets that far away. A spacecraft despatched on a hyperbolic path has kinetic energy greater than its potential energy and—even at infinity—it will still have some of this kinetic energy, and will still be moving at a finite velocity. To get anywhere worthwhile in a reasonable time, spaceprobes must be launched on hyperbolic trajectories with velocities in excess of escape velocity.

The table beneath illustrates a remarkable effect which arises because, in talking about escape from the Earth, we are dealing with energy.

Initial velocity from the surface of the Earth	final velocity after 'escape'
11km/s	0km/s
12km/s	5km/s
13km/s	7km/s
14km/s	8·5km/s

Taking the escape velocity of the Earth to be 11km/s, we find that if we launch a rocket with this value of initial velocity (v_i) then its final velocity after 'escape' (v_f) drops to zero—just as we would expect. If we were to launch it at 12km/s, just 1km/s greater than escape velocity, we would find the curious result that the final velocity came out at nearly 5km/s—i.e., an increase in launch velocity of 1km/s results in a final velocity increased by 5km/s. The answer to this apparent paradox lies in the nature of kinetic energy, which depends upon the *square* of the velocity; to obtain the final velocity we must look at the difference between the initial energy and the escape energy.*

---

*The initial kinetic energy $= \frac{1}{2}mv_i^2$, and the final kinetic energy $= \frac{1}{2}mv_f^2$. If escape velocity is denoted by v_e then the minimum energy to escape is $\frac{1}{2}mv_e^2$.

We then have that
$$\tfrac{1}{2}mv_f^2 = \tfrac{1}{2}mv_i^2 - \tfrac{1}{2}mv_e^2,$$
or, cancelling $\frac{1}{2}m$,
$$v_f^2 = v_i^2 - v_e^2.$$
Thus if $v_i = 12$km/s, $v_f^2 = 12^2 - 11^2 = 144 - 121 = 23$; therefore $v_f = 4.8$km/s.

Of course, escaping from the Earth is to some extent a case of 'out of the frying pan, into the fire', for the spacecraft then falls under the influence of the Sun's gravitational field, and must attain a higher velocity again if it is to exceed the solar escape velocity and move into interstellar space.

The Roads to the Planets

The Earth moves around the Sun at a mean distance of 149,600,000 kilometres at an average velocity of just under 30km/s, that being the value of circular velocity around the Sun at our distance. In fact, the orbit of the Earth is an ellipse, and the same goes for the other planets, too, but most of the planetary orbits are quite close to circular; exceptions are Mercury and Pluto, which have markedly elliptical paths. The outer planets (Mars to Pluto) move more slowly than the Earth and make their closest approach at a position called opposition (fig. 8), when Sun, Earth and planet lie in a straight line. The inner planets, Mercury and Venus, move faster than the Earth and make their closest approach at inferior conjunction, at which point they lie between Sun and Earth.

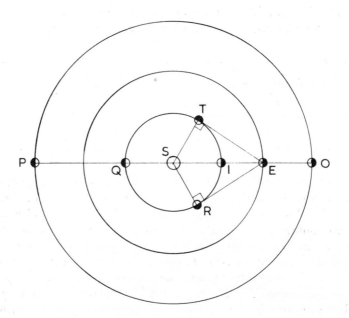

Fig 8 **The configurations of the planets.** For an outer planet (such as Mars), when the Earth is at E and the planet is at O, the planet is said to be at *opposition.* The point P represents *superior conjunction,* the planet being on the opposite side of the Sun from the Earth. For an inner planet (such as Venus), Q represents superior conjunction, I, *inferior conjunction,* and points R and T, *greatest elongation;* when the planet is at either of these two points, the angle in the sky between the Sun and the planet is at a maximum.

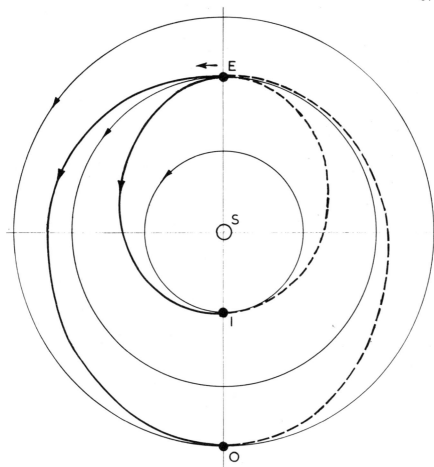

Fig 9 Hohmann transfer orbits to the planets. The orbit which requires the minimum expenditure of energy (and the minimum amount, therefore, of rocket fuel) is an ellipse which just touches the orbit of the Earth and the orbit of the target planet. To reach the outer planet (O) along such a path it is necessary to fire the rocket in the same direction as the Earth's motion, while to reach the inner planet (I) the rocket must be fired in the opposite direction to the Earth's motion so that its velocity subtracts from that of the Earth, and the rocket moves more slowly relative to the Sun than does the Earth.

At first glance it might appear that the logical thing to do if we wished to send a probe to Mars, say, would be to wait until Mars was approaching opposition and fire the rocket directly outwards, straight across the gap between the orbits of Earth and Mars, so taking the shortest route. Unfortunately, this is just what we cannot do. For a spacecraft to accomplish this it requires energy enough not only to climb directly away from the Sun (for as soon as it leaves the Earth the spacecraft will be dominated by the Sun's gravitational field) but also to cancel out the sideways motion which it possesses by virtue of the motion of its launch platform, the Earth. The Earth is moving at 30km/s, which is a considerable velocity, and most of us have encountered the problems raised by trying to jump off a moving vehicle travelling at a mere 30kph!

No rocket yet built has the ability to undertake such a flight; in any case, to do it that way would be extremely wasteful of fuel. There are subtler and more economical ways of solving the problem, as was first demonstrated in 1925 by the German town planner and architect, Walter Hohmann, who argued that the least expenditure of energy would be incurred if the rocket were fired in the same direction as the motion of the Earth, so *making use of* the Earth's orbital motion. After 'escape' the rocket would be moving away from the Earth in the same direction as the Earth and, relative to the Sun, would be in the position of a body which was initially in a circular orbit but has been given a higher velocity. Moving faster than the Earth, the spacecraft will follow an elliptical orbit having a perihelion distance (point of closest approach to the Sun) equal to the radius of the Earth's orbit and, hopefully, an aphelion distance (greatest distance from the Sun) equal to the radius of the target planet's orbit. A Hohmann transfer orbit (fig. 9) is an ellipse which just touches the orbit of the Earth and the orbit of the target planet (more generally, it is an ellipse which links any two orbits).

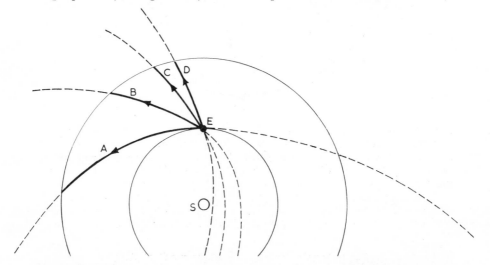

Fig 10 **Fast orbits to the planets.** The target planet may be reached in a shorter period of time than would be required for a Hohmann orbit (see fig. 9) at the cost of a higher expenditure of energy. Trajectory A is an ellipse relative to the Sun achieved by launching the rocket in the same direction as the Earth's motion but at a higher speed than that required for Hohmann transfer. B is a section of an ellipse where the launch velocity was angled to that of the Earth. C and D are hyperbolic paths relative to the Sun; to follow such paths the rocket must exceed the escape velocity of the Solar System by a substantial amount. ABCD represent a decreasing sequence of travel time but an increasing sequence of energy expenditure.

Any other interplanetary orbit will require a greater expenditure of energy; the Hohmann transfer orbit is the cheapest way to go. The penalty to be paid is in the time taken to complete the mission; Hohmann orbits require longer flight times than the more expensive 'fast' orbits (see fig. 10). For example, if we consider a flight to Mars (ignoring the elliptical nature of that planet's orbit) along a Hohmann orbit, the velocity of the probe relative to the Earth after 'escape' need be only 3km/s, but the probe will require nearly nine months to reach its target.

The launch velocity from the Earth's surface needs to exceed escape velocity by only 400 metres per second in order that the final velocity be 3km/s. By contrast, to follow the straight-line route 'across the gap' the final velocity of the probe relative to the Earth would need to be about 32km/s, and that would require an initial launch velocity three times greater than the required value for the Hohmann transfer. The economy of the latter speaks for itself. (Examples of Hohmann orbits are given in Table 2.)

The situation is complicated by the fact that the use of such orbits severely restricts the occasions on which launches may take place. For example (fig. 11), in order to reach Mars the spacecraft must be launched well before opposition, when Earth and Mars are in the correct relative positions. Encounter will take place after opposition when the spacecraft and Mars get to the same point, hopefully at the same time (if not, the mission controller is likely to find himself smartly

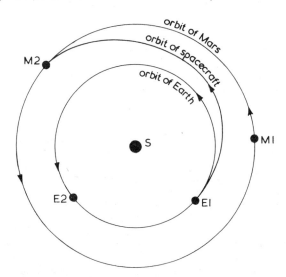

Fig 11 **Launch window for a voyage to Mars**. For a spacecraft to reach Mars by a Hohmann transfer orbit (see fig. 9)—because Mars moves more slowly than the Earth—the vehicle must be launched before opposition (when the Earth is at E1 and Mars at M1) in order to meet up with Mars at M2. By the time of encounter, the Earth will be well ahead of Mars, at E2.

placed on the unemployed register). The times during which launches can take place are known as 'launch windows'; in fact, existing rockets are capable of exceeding the minimum required velocities and so broadening these windows.

If a return journey along a Hohmann orbit is envisaged, then the time taken on the return journey will be the same as that required on

5
A photograph taken in September 1976 by Viking 2 of Mars' inner satellite, Phobos; the range is 880km and the resolution is such that the smallest object visible is about 40m across. The moon is seen to be heavily cratered, as had been expected, but there are also rather inexplicable striations and chains of smaller craters; where these occur on the Moon, Mars and Mercury they are thought to have been the result of secondary cratering from a major impact, but with a body the size of Phobos (about $13 \cdot 5 \times 10 \cdot 7 \times 9 \cdot 5$km), with a similarly low gravity, this seems unlikely. (*NASA.*)

6
The globular star cluster M13 (NGC 6205) in the constellation of Hercules. Our Galaxy possesses a halo of such clusters, each being of diameter between about five and fifty parsecs. (*Photograph from the Hale Observatories.*)

7
The North America Nebula in the constellation of Cygnus, an example of a gaseous nebula in which, astronomers believe, new stars are currently in the process of formation out of interstellar material. The brilliance of the nebula is owing to the radiation of such young, very hot stars. (*Photograph from the Hale Observatories.*)

8
The Great Spiral Galaxy in Andromeda (M31, NGC 224) with its two attendant satellite galaxies. The Andromeda Galaxy is the closest to our own—with the exception of the Milky Way's own satellite galaxies, the Large and Small Magellanic clouds and some minor systems—and, although somewhat larger, is very similar in appearance; it is about 670,000 parsecs distant. Telescopes can, at highest resolution, resolve individual stars in the Andromeda Galaxy (it should be borne in mind that all the *individual* stars in *this* picture belong to our own Galaxy), but the central hub is rather enigmatic since such resolution has not proved possible; this has led to the suggestion that galaxies have at their hearts some sort of 'superstar', although such a suggestion is highly speculative. (*Photograph from the Hale Observatories.*)

9
The launch, in July 1971, of Apollo 15 on a lunar landing mission. The astronauts aboard were David R. Scott, commander; Alfred M. Worden, command module pilot; and James B. Irwin, lunar module pilot. (*NASA.*)

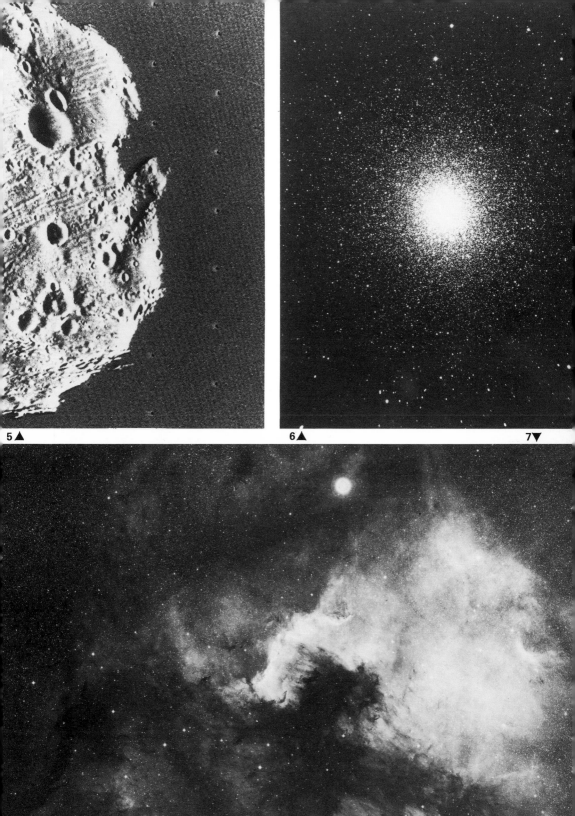

5▲   6▲   7▼

10 ◄

11 ▲

12 ▼

the outward journey, *but that return cannot be made immediately*; the space-craft—and its crew if it is manned—must await a suitable window to embark on the journey home. In the case of Mars the waiting time would be more than a year, so that the whole mission would involve a timescale of the order of three years.

TABLE 2

HOHMANN TRANSFER ORBITS

The velocity and time requirement to send a probe from the Earth to each planet in the Solar System is given below. (Earth and planets are assumed to be at their mean distances from the Sun.)

Planet	Flight time (one-way)	Required spacecraft velocity relative to Earth (km/s)	Launch velocity from Earth's surface (km/s)	Launch velocity minus escape velocity (km/s)
Mercury	107 days	7·5	13·5	2·3
Venus	146 days	2·5	11·5	0·3
Mars	260 days	2·9	11·6	0·4
Jupiter	2·7 years	8·8	14·2	3·0
Saturn	6·0 years	10·3	15·2	4·0
Uranus	16 years	11·3	15·9	4·7
Neptune	30 years	11·6	16·1	4·9
Pluto	45 years	11·8	16·3	5·1

To move inwards, closer to the Sun, it is necessary to fire the rocket in the opposite direction to the Earth's motion so that it ends up moving relative to the Sun more slowly than the Earth, and thus falls inwards on an elliptical path. Curiously enough, it is harder (in energy terms) to send a probe from the Earth to the Sun than to most of the neighbouring stars, and, although we have already despatched probes that will leave the Solar System altogether, we cannot send one directly to our own parent star. The reason is quite simple. To reach the Sun, a probe must, at the very least, wholly cancel any sideways (transverse) velocity; if it has any sideways motion it will follow an ellipse which

10
The launch of the Russian Soyuz craft. (*Novosti Press Agency.*)

11
An artist's impression of the Russian Salyut long-endurance orbital station, with a docking manoeuvre under way. (*Novosti Press Agency.*)

12
A photograph taken in June 1973 from the command module, during its final 'fly around' inspection, of the Skylab space station. (*NASA.*)

misses the Sun. Launching from the Earth, the Earth's transverse motion of 30km/s must be nullified by firing the rocket in the opposite direction at just this velocity. It would then be stationary relative to the Sun, and would begin to fall directly towards the Sun like a stone dropped from an exceedingly great height. The minimum velocity required after 'escape' is 30km/s. The solar escape velocity at the Earth's distance from the Sun is 42km/s but, as we are in any case moving at 30km/s, all we need to do to get away permanently from the Solar System is to add a further 12km/s after escape.

Interplanetary Billiards

On the face of it, it would seem that with existing rockets we are condemned to long flight times if we wish to reach the outermost planets. However, there is one ingenious trick in the space scientist's armoury which has already been used to accelerate spacecraft and to change their orbits *without any expenditure of fuel* apart from that required for minor manoeuvres.

The basic approach is to make use of the gravitational field of one or more planets to change the velocity of the spacecraft. If we were to send a spaceprobe out towards giant Jupiter with just sufficient velocity to get there or a little way beyond, then by the time the probe reached the vicinity of Jupiter's orbit it would be moving much more slowly than Jupiter itself. As we see in fig. 12, Jupiter catches up with the slow-moving probe and, as the spacecraft falls within the gravitational influence of that planet, so it is accelerated. Viewing the situation from Jupiter's point of view, the spacecraft approaches, accelerating as it does so, and (if it has been properly targeted) rushes past on a hyperbolic path; as it moves away again, it slows down until by the time it is a long way off, it is receding from Jupiter at just the same speed with which it began its initial approach. You may say, 'nothing has been gained', and so it may seem.

However, viewing the encounter from the Sun's point of view a different picture emerges. Relative to *Jupiter* the spacecraft is moving no faster than it was, but as a result of the encounter its direction has changed. The spacecraft is now moving in much the same direction as Jupiter itself, but whereas, before, the spacecraft speed relative to the Sun was less than that of Jupiter, it is now greater (the probe velocity relative to Jupiter is now added to that of Jupiter). In effect, the spacecraft has picked up additional velocity roughly equal to that of Jupiter in its orbit. The orbit of the probe will now be a new ellipse which extends much further from the Sun, or it may even be a hyperbola.

Marvellous. This truly seems to be a case of getting something for nothing. Usually there is a catch in such schemes, but there does not seem to be one here. In fact, there is one, but it is rather academic. If the spacecraft has picked up energy then something must have lost an equal amount of energy, and that 'something' is mighty Jupiter; in

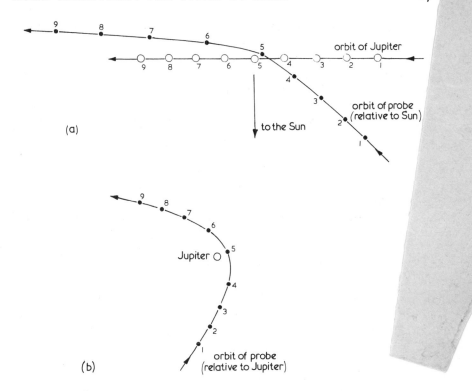

(a)

(b)

Fig 12 **Jupiter encounter.** In (*a*) Jupiter catches up with the slow-moving probe, and the outcome of the encounter is illustrated by considering the positions of Jupiter and the probe at nine instants (numbered 1 to 9). It may be seen that, *relative to the Sun*, the direction of motion of the probe has changed and its speed has clearly increased. (*b*) shows the situation seen from Jupiter's point of view (i.e., regarding Jupiter as stationary, which any Jovian is quite entitled to do) ; the spacecraft approaches Jupiter from the left (this is due to the fact that Jupiter has actually been catching up with the probe) and passes by on a hyperbolic path, receding in the direction shown. The combination of the deflection of the direction of the spacecraft relative to Jupiter, *and* Jupiter's orbital motion, results in the higher speed of the probe relative to the Sun.

accelerating the spacecraft Jupiter has been slowed down. The effect is immeasurably small, and we need have no fear of causing Jupiter to plunge into the Sun as a result of the encounter!

The amount of deflection and acceleration achieved may be varied by adjusting the velocity of approach and the distance of closest approach. For example, Pioneer 10 passed by at a range of 130,000 kilometres on 5 December, 1973, and was accelerated beyond the escape velocity of the Solar System, so becoming Man's first interstellar probe. It carries a plaque showing where it came from, just in case 'someone out there' should ever find it. By passing a probe ahead of Jupiter, the deflection may in principle be used to cancel the transverse velocity of the probe, allowing it to fall straight into the Sun. A long road for a short cut it may be, but Jupiter offers the means to visit our nearest star—as well as more distant ones.

The technique was also used to great effect with the planets Venus and Mercury by the Mariner 10 spacecraft, which flew close by Venus on 5 February, 1974, and was deflected by that planet into an orbit which took it past Mercury on 29 March, 1974, this mission resulting in the first close-up photographs of each of those planets.

Searching the Solar System

The first truly successful planetary probe (excluding those which went to the Moon) was Mariner 2, an American spacecraft launched on 27 August, 1962, which flew past Venus at a range of 35,000 kilometres on 14 December of that year, revealing that the planet is extremely hot and that its atmosphere is made up largely of the heavy gas carbon dioxide. Its sister-ship Mariner 4 made the first successful fly-by of Mars on 14 July, 1965, after a journey lasting 228 days. Since that time there has been a multitude of planetary missions, including fly-bys of Mercury, Venus, Mars and Jupiter and orbiters around, and landers on, the planets Venus and Mars. Progress has been spectacular and it is beyond the scope of this book to attempt a full account of the many inter-planetary probes, but all successful planetary missions up to the spring of 1978 are listed in Table 3.

It is no exaggeration to say that the planetary probes have revolu-tionized our understanding of the planets and their environments; many long-held theories have been overturned, and a wealth of informa-tion has been gained which quite simply could not have been obtained in any other way. That is not to say that the Earth-bound astronomers

TABLE 3

SUCCESSFUL PLANETARY PROBES

The table below includes only those probes which sent back significant quantities of results.

Planet	Spacecraft	Mission accomplished
Mercury	Mariner 10 (USA)	Fly-by, three times, 1974/75
Venus	Mariner 2 (USA)	Fly-by, 1962
	Mariner 5 (USA)	Fly-by, 1967
	Venera 4 (USSR)	atmospheric descent, 1967
	Veneras 5 and 6 (USSR)	atmospheric descent, 1969
	Venera 7 (USSR)	landing, 1970
	Venera 8 (USSR)	landing, 1972
	Mariner 10 (USA)	fly-by, 1974
	Veneras 9 and 10 (USSR)	landings, 1975
Mars	Mariner 4 (USA)	fly-by, 1965
	Mariners 6 and 7 (USA)	fly-bys, 1969
	Mariner 9 (USA)	orbiter, 1971
	Vikings 1 and 2 (USA)	orbiters/landings, 1976
Jupiter	Pioneer 10 (USA)	fly-by, 1973
	Pioneer 11 (USA)	fly-by, 1974

got everything wrong by any means. For example, when Mariner 10 flew past Mercury in March, 1974, and sent back photographs showing that Mercury has a crater-covered surface like the Moon, few were particularly surprised by this; but when Mariner 4 examined Mars in 1965 and showed it to be crater-covered, that did upset a lot of people at the time.

At the time of writing there are several probes in transit in various regions of the Solar System. Pioneer 11, which flew past Jupiter in December 1974, is currently *en route* for an encounter with Saturn in 1979, while its predecessor, Pioneer 10, should pass beyond the outermost planet in 1987, bound for interstellar space and an endless journey through the void. The two American Voyager probes, launched in August and September 1977, are embarked upon a veritable space odyssey which will last nearly a decade, taking in Jupiter and its satellites, Saturn and its giant moon Titan and—with luck—the planet Uranus with its newly discovered ring system.

It is highly likely that every planet in the Solar System will have been inspected by an unmanned probe by the turn of the century so that, within the first four or five decades of the Space Age, Man's explored territory will have encompassed the entire planetary system. Pioneer 11 will leave the Solar System (or at least the known boundary of the planetary system) just 30 years after the first Sputnik startled the world with its quaint 'bleep-bleep' signal as it hurtled across the sky. Had anyone seriously suggested thirty years earlier, in 1927, that within sixty years a manmade craft would cross the orbit of Pluto he would have been regarded at best as a crank and at worst as a lunatic. Such has been the pace of progress so far.

Nearer Home
Much has happened in near-Earth space over the past couple of decades, both in the area of manned spaceflight and in the satellite arena. The manned space programmes of the United States and the Soviet Union have to some extent run in parallel, and this was particularly so in the early stages. The Russians developed first the one-man Vostok, then the two/three-man Voskhod and the three-man Soyuz; while the equivalent stages in the American programme are represented by the Mercury, Gemini and Apollo spacecraft. The essential difference between the two approaches lay in the American lunar landing programme—a project on which the Soviet Union has not yet embarked.

Closer to the Earth, the first stages in establishing manned orbiting laboratories and workshops came with the launching on 19 April, 1971, of the first Soviet Salyut laboratory into an orbit some 200 kilometres high. The first crew, made up of Dobrovolsky, Volkov and Patsayev, docked their Soyuz 11 spacecraft with Salyut on 7 June, 1971, and spent 23 days aboard; tragically, due to a pressure leak in the Soyuz craft, the three men died during entry to the Earth's atmosphere.

Subsequent Salyuts have been visited by two-man crews for extended periods, the longest stay so far being achieved by Romanenko and Grechko who, in late 1977 through early 1978, remained aboard Salyut 6 for 96 days, thus establishing the space endurance record to date. A further key step on the road to maintaining men in space for prolonged periods was taken on this mission when the unmanned ferry vehicle, Progress 1, docked with further supplies for the crew on 23 January, 1978.

Skylab, the American orbiting laboratory constructed out of left-over Apollo hardware, entered Earth orbit on 14 May, 1973. Crews were ferried to and from it by means of Apollo-type spacecraft. With the Apollo craft attached, Skylab had an overall length of 36 metres, a mass of 90 tonnes and a habitable volume (in the orbital workshop, the principal component) of some 290 cubic metres, three times the equivalent volume of Salyut. Despite damage sustained during launching, which was effectively repaired by the first crew, Skylab proved to be an outstanding success, a total of 513 man-days in space being devoted to solar astronomy, Earth observations, astrophysics, the life sciences and various technological projects. The final crew remained for 84 days. The physiological results of importance to the long-term survival of Man in space appeared to indicate that Man did adapt quickly to a zero-gravity environment,* the only disturbing note being struck by the measurements that showed a steady decline in bone calcium, which showed no indication of halting during the 84-day period. This effect, if prolonged, would lead to brittle bones and obvious hazards for returning astronauts. Further, longer-term, research is clearly needed in this area, and it may prove to be necessary to simulate gravity on long-duration space missions.

Orbiting satellites have become a wholly unexceptional part of every-day existence now: communications satellites operating on a commercial basis link much of the globe with telephone and television channels; meteorological satellites provide detailed coverage of the Earth's weather systems, leading to improved weather forecasts and a better understanding of the factors which influence weather patterns. Earth resources can be monitored remotely by satellites such as the remarkable Landsat, the first of which was finally 'pensioned off' in 1978 after five and a half years of successful operation (its planned life expectancy was *one* year), which studied natural features, manmade structures, minerals, vegetation and oceans and has countless applications. Landsat data can be employed for such wide-ranging projects as measuring crop acreages, detecting offshore oil slicks or even locating potential earthquake zones.

Scientific satellites, meantime, investigate the upper atmosphere, the space environment, and distant astronomical sources emitting radiation

*For an explanation of the phenomenon of zero gravity, or 'weightlessness', see fig. 13.

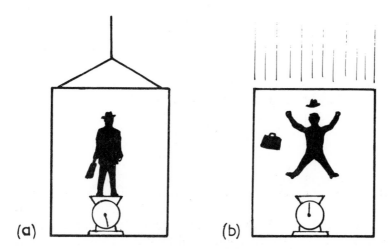

Fig 13 **Weightlessness.** (*a*) The person standing on the spring balance inside an enclosed box suspended in the Earth's gravitational field feels his normal weight. (*b*) If the box is released and allowed to fall freely, then all its contents will fall at the same rate and all impression of weight will vanish. Spacecraft coasting along in the gravitational field of the Earth, the Sun or other celestial body are in a state of free fall, and their occupants experience weightlessness.

which cannot be detected at ground level because it is unable to penetrate the atmosphere. Most of us, I am sure, are quite unaware of the scale of activity which is going on above our heads; well over four thousand pieces of manmade debris are floating around in orbit above the Earth. The military aspect is there too, of course, in the form of spy satellites and perhaps more deadly devices. On the other hand, the first steps have been taken towards international cooperation in space with the link-up of the Soviet Soyuz and American Apollo craft in orbit on 17 July, 1975. Let us hope that the future exploitation of space will lead to further cooperation, and that space will not become the new battlefield for rival ideologies.

We are already well into the era of space exploitation. Where we go from here is the subject of the next chapter.

5 Home Territory:
(ii) the way ahead

In the next few decades we may expect interesting developments in a number of areas: unmanned spaceprobes will extend their range, capabilities and variety, applications satellites will play an increasing rôle in our lives, men and women will become much more at home in space, space industry will grow in importance, and we will begin the process of the 'humanization' of our space environment. On a broader front, we should see the first manned planetary missions and the establishment of a lunar base, together—perhaps—with the construction of colonies *in* space. All of this is possible, and indeed most probable, provided that this prospect is not snatched away by the collapse of human civilization.

In the field of spaceprobes, the immediate future is fairly clear. Following the Venus Pioneers of 1978 (an orbiter, and a set of atmospheric probes), tentative plans by NASA, the United States Space Agency, include some of the missions described below. A lunar polar orbiter mission is one possibility for around 1980, the aim being to map the polar regions which have not so far received detailed coverage, and in which, it has been argued, there may be deposits of ice. A Mars rover seems a strong possibility for 1984. One or, possibly, two mobile robot vehicles would land on the surface of the red planet and traverse at least 100 kilometres during a year-long mission (the Soviet Union pioneered remote rover vehicles with their lunar vehicle Lunokhod 1 in 1970). Each rover, about the size of a large desk, would have loop wheels, stereo cameras and a manipulator arm (see Plate 30); the 100-kilogram scientific payload would include packages which could be deployed at interesting sites. With onboard computer systems, the rover would be able to manoeuvre independently of instructions from the Earth, avoid obstacles, and 'learn' routes· by itself and would carry a 'science alarm' to alert scientists on Earth immediately of any particularly important discoveries. Altogether, it will be a most impressive robot which, because it will not have to wait for instructions as to what to do next after each manoeuvre (the signal travel time from Earth to Mars—one-way—varies between 3 and 22 minutes depending on the relative positions of the two planets), will be far more effective than if under Earth control. A more basic rover may be launched earlier.

The possibility of an automatic Mars sample return mission is also being debated, with a possible mission about the same date as that scheduled for the rover. The technique was first used in 1970 by the Soviet probe Luna 16, which brought back 100 grams of material from the Moon's Sea of Fertility. The technical difficulty is very high, but NASA studies indicate the feasibility of a mission capable of returning a few hundred grams of Martian material to Earth. Such a mission raises broader issues. Whereas there was little doubt that lunar samples would be sterile, there is as yet no certainty that Mars does not support some very basic and unfamiliar lifeform which might be in bacterial form. It is just possible that such bacteria might find the environment of the Earth very much to their liking and, if let loose, might flourish to the detriment of indigenous lifeforms in general and ourselves in particular. Although the chance of terrestrial contamination may be remote, I do not believe the risk should be taken. Since we shall soon have the facilities to make such analyses in orbital laboratories, I am convinced that a sample return mission should await that kind of development.

The much more hostile conditions on Venus render a rover mission unlikely for some time. The next dramatic development is more likely to be an atmospheric buoyant probe, floating beneath a balloon in the upper atmosphere of the planet. Even at an altitude approaching 60 kilometres, a 6-metre hydrogen-filled balloon ought to be able to support a payload of about 80 kilograms which could sample Venusian meteorology for a period of weeks or even months.

Giant Jupiter will undoubtedly attract further attention, and an orbiting probe is currently under study at NASA's Jet Propulsion Laboratory. Tentatively scheduled for a 1982 launch—as the first planetary probe to be launched from the shuttle vehicle described later in this chapter—the orbiter would arrive at Jupiter 2.9 years later and release a probe which would descend rapidly through the planet's extensive atmosphere radioing back data to the orbiter which would, in turn, relay these findings to the Earth. The atmospheric probe would survive at most half an hour after entering the atmosphere, but the orbiter will be scheduled for some twenty months' operation. (Plates 28 and 29.)

Other wide-ranging missions under study include asteroid sampling probes and a comet probe, possibly scheduled to meet up with Halley's comet on its next return in 1986. By the early part of the twenty-first century we can confidently expect to have detailed knowledge of the atmospheres, surfaces and structures of all the planets, and detailed information concerning the Solar System's minor members too.

Most of the missions mentioned above are within the capabilities of existing rockets, but they would be greatly facilitated by superior propulsion systems. Two approaches which have been under study and development for some time are the nuclear rocket and the ion rocket. As we saw in the previous chapter, the rocket is basically a mechanism

which expels mass ('reaction mass') in one direction so as to be propelled
by reaction in the opposite direction, and the liquid-fuelled chemical
rocket of today expels hot gas produced by chemical reactions between
fuel and oxidant. The nuclear rocket is not so very different. Essentially
it requires a nuclear fission reactor (not unlike the reactors found in
nuclear power stations) to supply a great deal of heat to the fuel
(typically liquid hydrogen) which passes through and around the core
of the reactor. The fuel is vapourized and—because of the tremendous
heat supply—emerges from the rocket nozzle at a much higher exhaust
velocity than we would expect to attain from a conventional rocket
(fig. 14).

Studies of nuclear rocket systems began in the United States in about
1958 and led to the development of NERVA (Nuclear Engine for
Rocket Vehicle Application) in which an 1100Mw nuclear reactor
heated a working fluid of liquid hydrogen and generated a thrust of
about 24 tonnes. The exhaust velocities attained were several times
greater than those available from chemical rockets. There are considerable technical difficulties with such a device and as yet there are no
immediate plans to use such an engine for interplanetary missions.

The principle of the ion rocket is illustrated in fig. 15. Charged
atomic particles (massive positive ions; i.e., atoms stripped of some or
all of their encircling electrons) are accelerated by an onboard electric
field to exhaust velocities far in excess of the values associated with
conventional rockets. The amount of thrust such motors generate is
small, so they are incapable of lifting themselves off the ground, but in
space their ability to sustain this thrust for long periods allows, in
principle, very high velocities to be achieved. In practice, a gaseous
propellant such as mercury or caesium is passed into a discharge
chamber where it becomes ionized, the electric field between two grids
accelerating the positive ions. It is important to eject an equal amount

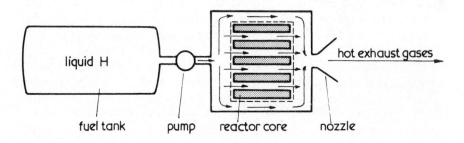

Fig 14 **The nuclear rocket.** A 'working fluid' such as liquid hydrogen is
pumped from the fuel tank around and through the core of a nuclear fission
reactor and in this process is severely heated. The resulting very hot gas
streams out through the nozzle at an exhaust velocity well in excess of that
attainable by chemically fuelled rockets.

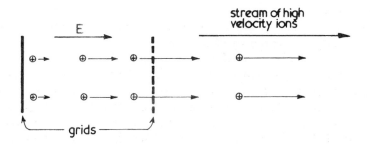

Fig 15 **Principle of the ion rocket.** Positively charged particles (ions) produced in a discharge chamber are accelerated by an electric field between two grids. The resulting exhaust velocity may be ten or twenty times greater than what can be obtained in chemical rockets. An important point is that the spacecraft must be kept electrically neutral by arranging to 'dump' an equal quantity of negative ions (electrons).

of negative ions (electrons), too, so as to keep the spacecraft electrically neutral—the effect of a highly negatively charged spacecraft entering the atmosphere would be spectacular, to say the least.

One scheme under active development by NASA is the Solar Electric Propulsion System (SEPS) which utilizes solar energy collected by giant panels of solar cells and converted into the electrical power required to achieve ionization and sustain the accelerating electric field. The SEPS system will probably consist of a cluster of eight to ten engines each using liquid mercury propellant with an exhaust velocity of 30 to 50km/s, and each generating only 0.2 kilograms of thrust! This tiny thrust may be applied for months, or even years; for a typical mission the acceleration per *day* would be 86m/s, but after 100 days the velocity relative to the Earth would have increased by 8.6km/s. Given an initial launch velocity from Earth orbit of 16km/s, the velocity of the SEPS vehicle would by then be 24.6km/s, or about 88,000kph. After 500 days' thrust the velocity would be something like 60km/s (210,000kph). The potential of this system is very impressive, as indeed will be its physical size—the 'wings', made up of solar panels, will extend over a distance of 150 metres, which is longer than a football ground.

One of the first missions scheduled for this system is an ecounter with Halley's comet on its next return in 1986; as this comet follows a retrograde orbit (i.e., it moves in the opposite direction to the planets in their orbits) very high velocities are required if a probe is to match orbits and fly along in parallel with it. The SEPS is capable of achieving this. Other possibilities include a heavy-payload mission to Saturn, a round trip to Mars and a fascinating scheme for sampling the surfaces of asteroids by trailing long sticky nylon ropes over their surfaces without having to achieve a landing. This concept of a cosmic fisherman is illustrated in Plate 32.

The Solar Sail

Of the various modes of interplanetary travel currently under consideration, by far the most romantic and aesthetically pleasing must be the solar sail. Radiation from the Sun exerts a tiny but finite pressure on any surface on which it falls. Although, under everyday circumstances, the pressure is too small to be noticed, it will give rise to a significant accelerating force on a body of sufficiently low mass and sufficiently high surface area, as was dramatically demonstrated in the evolution of the orbit of the US balloon satellite Echo 1, launched in 1960. Radiation pressure caused the perigee altitude to vary by some 500 kilometres.

The possibility of using solar radiation to propel a spacecraft was first discussed in accounts written in 1921 and 1924 by the Soviet scientists K. Tsiolkovskii and F. A. Tsander, and the principle is now receiving serious attention. Is it really possible that one day soon we may see sailing craft plying their way around the Solar System like the sailing ships of old?

Light may be considered as a stream of particles, photons, each carrying a small quantity of momentum. As a photon strikes a reflective surface it transfers momentum to that surface; a steady stream of photons exerts a steady pressure on a surface, and if we have a 'sail' sufficiently large and light this pressure may be used to drive a spacecraft. The acceleration obtained depends on a number of factors: the combined mass of sail and spacecraft, the reflectivity of the sail, the angle which the sail makes with the rays of sunlight, and the distance from the Sun (radiation pressure diminishes as the square of distance).

If we were to construct a sail with a surface area of one square kilometre, how effective would it be? At the Earth's distance from the Sun, radiation pressure amounts to five millionths of a Newton* per square metre, and over the area of the sail, one million square metres, the net accelerating force would be some 5 Newtons. As it would be quite feasible to build a sail/spacecraft combination having a total mass of 5 tonnes, this force would accelerate the craft at a rate of 1 *millimetre* per second per second (about 0.0001*g*). Although this may seem paltry, the important thing is that it can be sustained for very long periods of time. After 1000 seconds the spacecraft would have accelerated by 1 metre per second, while after 1,000,000 seconds—about 11½ days— the extra velocity gained would be 1 *kilometre* per second. Without expenditure of fuel the craft would have picked up an additional 3600kph. Surely such a means of getting something for nothing should not be ignored!

Since sunlight travels directly away from the Sun it is not difficult to visualize the way in which a solar sail may move outwards from the

---

*One Newton is the force required to accelerate one kilogram at a rate of one metre per second per second; i.e., to produce an acceleration of about 0.1*g*.

Sun, but it is less easy to see how such a sail could manoeuvre a space-craft closer in. Sailing enthusiasts will be familiar with the technique of sailing to windward by a series of 'tacks' (or 'boards') made at an angle of about 45° to the wind direction. The solar sailor can also progress 'to windward', but the analogy with a sailing yacht should not be taken too far, for a key factor in the sail's behaviour is the gravitational attraction of the Sun.

The force on a flat solar sail is exerted at right angles to its surface (fig. 16). If the sail is tilted to the direction of the Sun, we may say the resultant force has two components, a radial component directed away from the Sun and a transverse component exerted at right angles

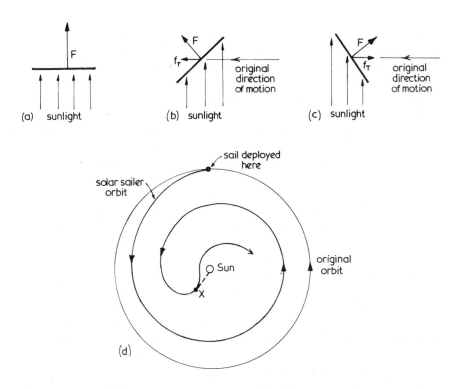

Fig 16 **The Solar Sail.** The pressure of sunlight acts perpendicularly to the surface of the sail. In (*a*) the sail is at right-angles to the direction of the Sun and the resultant force (*F*) acts radially outwards from the Sun. In (*b*) the sail is tilted so that a component of F (f_t) acts in the direction of motion of the spacecraft, so increasing its velocity, while at (*c*) the sail is tilted so as to produce a component of force (f_t) acting *against* its motion, so reducing its speed.

In (*d*) the result of maintaining the sail in configuration (*c*) is to reduce the transverse ('sideways') velocity of the spacecraft to zero at the point X. If the sail is then furled, the craft will fall into the Sun. If it is kept in action, the spacecraft will begin to pick up speed in the opposite direction to its original velocity and will spiral outwards again but in a retrograde direction.

to this direction. The greater the angle of tilt the greater the ratio of the transverse component to the radial component, but, as the angle of tilt increases, the effective area of sail presented to the Sun's rays diminishes; at an angle of 90° the sail is edge-on to the solar radiation and the driving force is zero. If our solar sail were initially in the same orbit as the Earth it would have a transverse velocity of 30km/s; if the sail were tilted so as to exert a transverse acceleration in the direction opposite to the motion of the spacecraft then its transverse velocity would be reduced and it would spiral in towards the Sun. In practically all situations, the sail may be tilted to exert a force against the direction of motion.

Over a period of time the transverse motion of the spacecraft could be reduced to zero and, if the sail were then furled, the craft ultimately would fall straight into the Sun. Alternatively, if the sail were kept set as before, radiation would build up its velocity in the opposite direction to the original motion and the craft would begin to spiral out on a retrograde orbit. By means of the appropriate balance of gravitational attraction and radiation pressure, a wide variety of maneouvres would be possible.

NASA's Jet Propulsion Laboratory is actively considering a number of proposals, such as the 'Yankee Clipper', which would utilize a square sail 800 metres along each edge made of aluminized plastic only 0.0025 millimetres thick; and the 'heliogyro', a spinning solar 'windmill' consisting of twelve thin reflective blades each 7.4 *kilometres* long by 8 metres wide. Although the solar sail was rejected in favour of the SEPS for the proposed Halley's comet mission, I have no doubt that this elegant scheme will one day see the light of day.

The solar sail cannot be considered for interstellar missions, for solar radiation pressure diminishes rapidly with distance;* at a range of ten astronomical units (about the distance of Saturn) it is only 1% of what is available at the Earth's distance. Closer in, of course, the propulsive force is greatly increased, but a solar sail would have to be careful to avoid the fate of the legendary Icarus if it approached the Sun too closely.

It seems highly likely that solar sails will propel scientific payloads in the future and, just as there are signs today of a revival of commercial sail on the Earth's oceans, so we may see sail-power propelling interplanetary cargo vessels. It may not be too far-fetched to suppose that future generations may participate in the 'Single-handed Mars Race' or the 'Round the Sun Race' in fleets of racing 'yachts'. In some ways such voyages would be easier than ocean races, the wind strength and direction being wholly predictable, but there would still be considerable skill involved in trimming sails to best advantage, and in the navigation!

---

*But see page 134 for a description of a laser-powered interstellar 'sail'.

Shuttles and Spacelabs

The development of a multitude of artificial satellites will doubtless proceed apace, but the next major step forward in manned spaceflight and in the utilization of near-Earth space will come with the Shuttle, the re-usable launch vehicle currently being test-flown in the United States. One of the major factors in the cost of space missions is that the launch vehicle is thrown away on each mission; it is used once and cannot be recovered, and, with something like a Saturn 1B costing in the region of $100,000,000, this is a major item. Each Shuttle orbiter vehicle will cost in the neighbourhood of $300,000,000, but will be designed to be re-used about one hundred times before requiring major overhaul. The hope is that the Shuttle system will reduce the cost of launching or retrieving orbiting payloads by a factor between 5 and 10.

The Shuttle will take the form of an orbiter mounted 'piggy-back' on a large expendable propellant tank together with two strap-on re-usable solid fuel boosters; the whole configuration will be launched vertically from a conventional launch pad (Plate 15), the boosters will return by parachute, and the orbiter itself will glide to a conventional landing on a runway. The orbiter looks like a delta-wing aircraft comparable in size with a commercial airliner such as the DC-9, the overall length being 37 metres and the wingspan 24 metres. Much of the body is taken up by the cargo bay, which measures 18 metres long by 4.5 metres wide and has a payload capacity of 29.5 tonnes. Piloted by a crew of two, the orbiter has space for one or two mission specialists, but—exceptionally—up to four mission specialists can be carried for missions lasting from a week to thirty days.

The scope of the Shuttle is considerable. The cargo bay can carry a variety of different payloads (say a batch of satellites) at the same time, and not only can satellites be carried up and thoroughly tested before being left in orbit, but malfunctioning satellites can be repaired in orbit or retrieved and brought back to Earth. An additional potential saving with the Shuttle is revealed in a NASA study of past satellite failures, which indicates that, out of 131 failures, 78 were due to launch problems and 53 to spacecraft anomalies; most of these failures could have been avoided had the Shuttle been available.

The Shuttle has its limitations, too. It is not designed to remain in orbit longer than 30 days and, as a result, does not contain the relatively palatial accommodation of Skylab; it will not be able to contribute directly to the study of the long-term effects of weightlessness. The maximum altitude to which it can go will be about 800 kilometres. Payloads which need to go higher than this (for example, geosynchronous satellites*) will be transported by the space tug, a retrievable

---

*A geosynchronous satellite is one which is placed in orbit 36,000 kilometres above the Equator so that its orbital period is equal to the rotation period of the Earth. Such a satellite remains above a fixed point on the Earth's surface; this is particularly useful for communications purposes.

unmanned rocket vehicle which—in effect—acts as the 'third stage' of the Shuttle.

The Shuttle, for the first time, opens the door for non-astronaut-trained persons to fly in space; reasonable physical and mental fitness should be all that is demanded. Among the most interesting of potential payloads will be Spacelab, a manned laboratory designed to fit in the Shuttle cargo bay, which is currently being developed by the European

13
Thomas P. Stafford (USA) and Alexei A. Leonov (USSR) during the historic Apollo–Soyuz link-up of July 1975. The view is from the Soyuz craft into the docking module. The two astronauts concerned were the commanders of the vessels engaged in this unique maneouvre during which, for two days, the Soyuz and Apollo craft remained docked together and the astronauts visited each other. (*NASA.*)

overleaf
The arrival of a Space Ark at a planet of another star. The Ark is an O'Neill cylinder to which have been attached a propulsion unit and additional shielding to cope with erosion by particles in interstellar and planetary-system space. Ahead of the Ark travels a mobile shield to catch the impact of more massive objects, particularly as the Ark leaves the Solar System and, centuries later, as it enters the target system.

The cylinder is about 20km long and about 4km in diameter. It has three reflectors, hinged at their bases, to reflect light into the Ark through long windows when it is in the vicinity of a star; in interstellar space these reflectors are closed up and the energy requirements of the Ark are met by thermonuclear reactors.

The Ark has recently arrived in this planetary system and is now in orbit several tens of thousands of kilometres above an Earth-type planet. The centre of the system is an orange star which is rather cooler than our Sun; the planet lies significantly closer to the star than does the Earth to the Sun, and so the star appears several times larger than does the Sun in our sky. At the upper right can be seen one of the planet's moons; it is small, irregular and rocky, rather like Phobos, the Martian satellite, and it is intended that the colonists will utilize its material to build further space facilities before establishing ground-based habitations.

The drive unit of the Ark is still attached and is being checked out by the occupants of a small space tug. An atmospheric shuttle is being tested, while the mobile precursor shield is being manoeuvred away to be dis-mantled and used for other purposes.

In the foreground is a beacon satellite which has been established in orbit about the planet to signify that initial steps in terraforming the planet have been instituted by the crew of one of a new generation of fast starships constructed several generations after the departure of the Ark; nevertheless, the superior speed of the fast starship has enabled its occupants to have already checked out the Ark's target system.

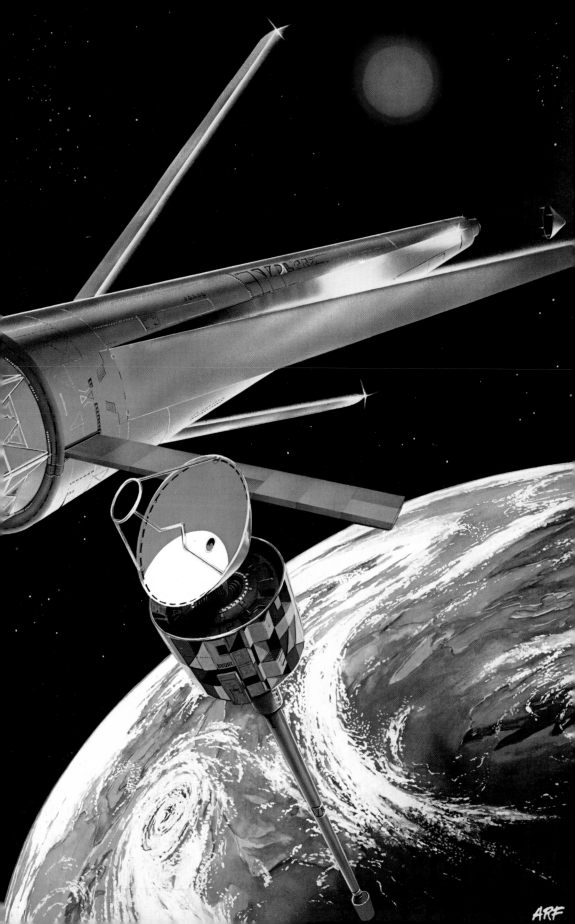

Space Agency in conjunction with NASA. It consists of two basic elements: a pressurized manned section in which scientists can work in a 'shirtsleeves' environment, and an external platform or 'pallet' to carry instruments which require direct exposure to space. With access available to permanent zero gravity, ultra-high vacuum and the full spectrum of natural radiations, a wide variety of Spacelab missions is possible, including work in astronomy, Earth surveys, the life sciences and industrial technology. In connection with the last item, space has already proved to be a most suitable environment for the manufacture of new kinds of alloys, ultra-pure glass (for optical, electronic and laser applications) and for the growth of large crystals of high purity for use in electronics.

From the astronomer's point of view perhaps the most interesting proposal is the Space Telescope, scheduled for the mid-1980s. Of 2.4-metre aperture, this telescope should be able to detect galaxies and stars 100 times fainter than the best so far attained with ground-based telescopes, and will greatly enhance the range to which we can see in the Universe.

The Shuttle Orbiter is now at an advanced stage of testing. Orbiter 101, the first Shuttle, was put on public display on 17 September, 1976, when it was named *Enterprise* (*Star-Trek* fans please note). Since then it has undergone a variety of tests including being carried 'piggy-back' on a Boeing 747 Jumbo jet for a number of linked flights before being released to glide home independently for the first time on 12 August, astronauts Haise and Fullerton in the hot seats. First orbital flights should begin in 1979/80.

The Shuttle is an essential element in bringing the space environment into Man's compass, but larger-scale space-stations too will be vital, stations which can sustain substantial numbers of people for months or even years. A variety of systems is under consideration, including modular structures which can be added to as time goes on. It seems quite likely that the USSR will develop such a system using the Salyut laboratory with multiple docking ports to which a number of Soyuz craft can be joined. A first stage in this process was successfully achieved in January, 1978, when Soyuz 26—already docked with Salyut 6—was joined by Soyuz 27 for a period of five days.

14
The second man to stand on the Moon: Edwin E. Aldrin, Jr. on the 20th of July 1969, follows his commander, Neil A. Armstrong, down the steps of the lunar module ladder as he prepares to walk on the surface of the Moon. (*NASA.*)

Colonies in Space

The most exciting and ambitious scheme which is being debated at the moment is the building of large self-sufficient human colonies in space. Enthusiasm for this kind of project is rapidly gaining momentum, and serious design studies indicate the technical feasibility of building such structures within the next forty to fifty years.

The concept is not new, and many of the ideas which seem likely to be incorporated into such colonies when and if they are constructed were discussed some eighty years ago by that most remarkable of space pioneers, Konstantin Tsiolkovskii, in his novel *Beyond the Planet Earth* (eventually published in 1920). In the nineteen-twenties, J. D. Bernal wrote convincingly of the possibility of space-borne artificial habitats for Man.

In popular literature the idea of colonizing other planets has tended to dominate, but it must be admitted that man-made biospheres would have many advantages over planets, particularly in terms of the control which could be exerted over these environments in ways which must remain impossible on a planetary surface. One classic example is gravity. This is determined by the mass and size of the planet, and there is nothing a surface inhabitant can do about that—his weight is determined by the planetary attraction. If the gravitational field is too strong, moving about, or even standing up, will require far too much effort (and heart failure is likely to ensue); if the field is too weak then muscles and the human frame may become seriously weakened with disastrous consequences if a return to Earth is contemplated. On a manmade colony, artificial 'gravity' can be generated by the simple expedient of spinning the colony. Centrifugal force* pressing objects against the rim of the rotating structure gives an impression of 'weight' indistinguishable from that experienced in a gravitational field and the strength of 'gravity' may be adjusted to the desired level by adjusting the spin rate. Probably it would be most convenient to have artificial gravity equal to the terrestrial value.

The idea of space colonization has begun to be treated seriously only

---

*Centrifugal force is the apparent force experienced by, for example, the occupants of a sharply cornering car; or, with a spinning weight on the end of a piece of string, it is equal and opposite to the force exerted on the weight by the string in order to keep it in circular motion. It arises because of the inertia (resistance to acceleration) of all massive bodies. According to Newton's First Law of Motion a body continues to move in a straight line at uniform velocity unless acted upon by a force. A person contained inside a spinning cylinder is prevented from carrying on in a straight line by the wall of the cylinder and therefore feels an apparent force pressing him against that wall; the force experienced is equal to his mass multiplied by the acceleration to which he is subjected. The sensation of weight on Earth is equal to one's mass multiplied by the acceleration due to gravity; on a rotating colony one's 'weight' would likewise be equal to one's mass times the acceleration experienced. The value of the acceleration depends upon the rate of rotation (angular velocity) and the radius of the circle being traced out. The faster the spin rate, the greater the acceleration and the greater the feeling of weight; it follows, too, that closer to the axis of rotation the force diminishes and at the axis itself it is zero (see fig. 17).

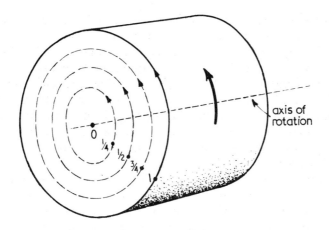

Fig 17 **Artificial gravity in a rotating cylinder.** The cylinder is rotating in the direction of the arrow, the resulting centrifugal force experienced by an observer on its inside surface being equal in magnitude to his normal weight. At the axis of rotation there is zero gravity, and at distances of $\frac{1}{4}$, $\frac{1}{2}$ and $\frac{3}{4}$ of the cylinder's radius (measured out from the axis) the effective 'weight' experienced is, respectively, $\frac{1}{4}$, $\frac{1}{2}$ and $\frac{3}{4}$ of normal.

in the last few years, and much of the drive behind this new approach stems from the work and infectious enthusiasm of Professor Gerard K. O'Neill of Princeton University, who gained academic respectability for the subject with the publication in 1974 of papers in the reputable scientific journals *Nature** and *Physics Today*.† Since that time, developments have been rapid. A ten-week summer study on space colonization was undertaken in 1975 jointly by NASA's Ames Research Center and Stanford University, and this proved highly productive in specifying the practical requirements of colonies and their structures. This has been followed by further NASA and university studies, and, as scientists, engineers and economists begin to tackle the nuts-and-bolts problems, the feeling is growing that this is an idea 'whose time has come'.

What conditions must be met by a satisfactory space colony? It must be able to sustain a pleasant, comfortable and safe environment in which a large number of people would *wish* to live, an environment which is self-sufficient and in which recycling is the order of the day. It must possess a powerful, cheap and wholly reliable energy source; and, of course, such a source is available in the form of the Sun since, in space, solar radiation in all its forms is available without interruption. The colony should have natural lighting (sunlight again) but must also be shielded from the ill effects of radiation—solar ultraviolet, X-rays and gamma rays, particles from solar flares, cosmic rays—and this shielding is likely to be the major mass requirement in the structure.

*G. K. O'Neill, 'Colonisation at Lagrangia', *Nature*, August 23, 1974.
†G. K. O'Neill, 'The Colonisation of Space', *Physics Today*, September, 1974.

For the first generations of colonists at least it will be desirable to create an environment as closely resembling that of the Earth as possible, for adjusting to the idea of living in a manmade habitat in the depths of space will be hard enough without being deprived of all the natural sights and sounds of the Earth. A feeling of spaciousness will be essential to avoid claustrophobia, so the colony cannot be too small or compartmentalized. The colony will probably be spun at a suitable rate to simulate Earth gravity at the rim, but any level of 'gravity' from normal down to zero may be experienced simply by moving to the appropriate part of the structure.

Many suggestions have been made as to where such colonies should be located, the most popular view being that they should be placed at the same distance as the Moon but 60° *behind* our natural satellite, at a point known as L-5. 'L-5' has a dramatic ring to it, and is the jargon abbreviation for 'the fifth Lagrangian point of the Earth–Moon system'. The French mathematician, J. L. Lagrange (1736–1813) showed that there were five stable points (libration points) at which a third body could be placed within the gravitational field of two massive bodies, such that the relative positions of the three bodies would remain fixed (fig. 18). Of particular interest are the points L-4 and L-5, where the three bodies make up an equilateral triangle, and, although in the Earth–Moon system the situation is complicated by the gravitational attraction of the Sun, it has been shown that any object placed near

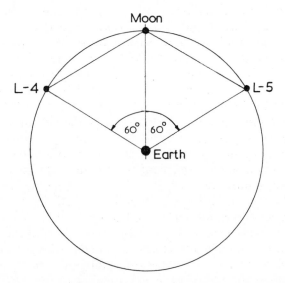

Fig 18 **The fourth and fifth Lagrangian points of the Earth–Moon system.** L–4 (the fourth Lagrangian point) lies ahead of the Moon, making an equilateral triangle with Earth and Moon; L–5 lies at a similar position behind the Moon. Objects placed at either of these points will stay in the vicinity of these positions, and will revolve around the Earth in the same period of time as does the Moon.

L-4 or L-5 would slowly circle about the libration point in a period of 89 days. In the vicinity of either point there is ample room to instal thousands of separate colonies, slowly drifting around.

Why put the colonies so far out? The argument in favour of the Lagrange points hinges on the supposition that practically all the material required to build the colonies will be mined on the Moon because *it will be far cheaper* to do so than to transport materials from the Earth to L-5 or even to a close-Earth orbit. This arises because, as we saw earlier, the gravitational 'well' of the Earth is much deeper than that of the Moon, and it requires far more energy to get out of it. The escape velocity of the Earth is 11.2km/s and that of the Moon is only 2.4km/s; since kinetic energy depends on the square of velocity it follows that something like twenty times as much kinetic energy has to be supplied to each kilogram of payload to get it away from the Earth compared to throwing it off the Moon. The Moon's lack of atmosphere is another factor and, of course, mining material from the surface of our satellite would have no effect on the environment of the Earth.

An early proposal by O'Neill for the first colony to house some 10,000 inhabitants was a pair of rotating cylinders each one kilometre long by 100 metres wide, but his current plan for Island One—as the first colony has come to be known—is to use a sphere some 460 metres in diameter which, rotating twice per minute, would generate a feeling of normal gravity at its equator. One reason for selecting the sphere (known as a 'Bernal Sphere' in honour of J. D. Bernal) is that it has the minimum surface area for the maximum internal volume, so minimizing the amount of radiation shielding required. The shell of the colony would be made of aluminium (plentiful in lunar rocks) surrounded by massive shielding made of raw lunar material or the slag left over after the material has been processed to release its metallic content. It has been estimated that the amount of mass required for the structure would be some 200,000 tonnes, that soil and internal structures would take up a further 400,000 tonnes, while the shielding would require about 3,000,000 tonnes, making a grand total of 3.6 million tonnes of lunar material to be transported to L-5.

Within the sphere would be located luxury apartments, soil, and grass and normal terrestrial vegetation growing under ideal conditions. Most of the habitations would be located in the equatorial regions within, say, 45° 'latitude' of the 'equator'; at these latitudes gravity would be reduced to 70% of normal, and at higher latitudes it would diminish further, reaching zero at the axis of rotation (the 'poles'). Low-gravity activities would be carried out near the axis, and one delightful possibility which has been suggested is that manpowered flight could take place in this region. To climb out of the equatorial valley towards the hub would be a most pleasant experience, since the effort required would diminish with altitude—quite the opposite experience to that encountered in climbing terrestrial hills.

Sunlight would be reflected into the colony through giant windows at the poles, and by a suitable alignment of mirrors could be made to enter at a fixed angle. Agricultural areas where crops could be forced on at optimum rates would be attached externally to the sphere, and access to these and to docking facilities would be achieved along the zero-gravity corridor along the axis. Light industry might be undertaken within the colony itself, but heavy industry, or processes requiring zero-g, high vacuum or high radiation levels would be carried out in units separate from the main structure.

The 1975 Ames Summer Study preferred the concept of a torus, or wheel-shaped colony (Plate 39), which would allow a feeling of normal gravity to be generated at the rim at a lower spin rate, and would allow long lines of sight from any point in the torus before the 'land' is lost behind the curvature of the wheel. It would also eliminate the direct awareness of having houses and people 'directly above your head', which would be inevitable in the Bernal Sphere type of colony.* In either case, a most pleasant environment could be created, free from terrestrial pests, and in which the inhabitants could get around everywhere on foot. A varied and pleasant diet should be no problem either, and there would be no need for the colonists to exist on soya stew if they didn't want to. Water would be extracted from animal and human wastes, and the entire ecosystem would be closed and as efficient as it is possible to be.

The chemical composition of the lunar rock (20–30% metals, 40% oxygen) seems ideally suited to the construction of colonies and the supplying of their environments, but how can it be mined and transported to Lagrangia? The necessary prerequisites for the operation would be as follows. A lunar mining colony would have to be set up, together with a basic 'construction shack' at L-5 itself, and to establish these—on current estimates—would require several tens of thousands of tonnes of material to be transported initially from the Earth. As a follow-up to the Shuttle, NASA is working on the design of a Heavy-Lift Launch Vehicle (HLLV) capable of lifting some 72 tonnes at a time into low Earth orbit. Material would then be transferred to the required location by some kind of tugboat, possibly powered by the SEPS technique described earlier in this chapter.

The tentative schedule for building the first colony envisages a 23-year operation with actual construction work on the colony itself not getting under way until about year fifteen. Most of the material (3.6 million tonnes) would have to be transported from the Moon in a period of six to eight years; i.e., at a rate of about 500,000 tonnes per

---

*It is curious to recall that rather more than 100 years ago Cyrus Teed published a book setting out his view that the Earth's surface is the *inside* of a hollow sphere, Australia being above the heads of Europeans and the Sun at the centre of the sphere; the density of air was supposed to prevent our seeing the antipodes. One is tempted to wonder—albeit briefly!—whether the Earth itself is not just a giant Bernal Sphere.

year. The key to this particular transportation operation lies in the low gravitational field of the Moon and its lack of atmosphere, the solution proposed being strongly reminiscent of Jules Verne's giant cannon, the 'Columbiad'. The lunar equivalent is the mass driver, which throws material at lunar escape velocity to a collecting site where it is gathered by the mass catcher in a kind of cosmic equivalent of baseball. The proposed driver uses a series of magnetically suspended buckets which are accelerated along a ten-kilometre track by means of superconducting magnets which build up the velocity at a rate of about $30g$. Finally the buckets are decelerated and the payload carries on; the principle is magnificently simple.

Material would then be transported to the colony site by means of a bulk carrier which could well be propelled by a variation of the mass driver mechanism—some of the payload being accelerated and expelled by an onboard accelerator, reaction pushing the spacecraft in the opposite direction. The carrier might take months to reach L-5, but with a continuous operation this time lag would not be important. The mining operation back on the Moon would not be a massive one—the entire mass required would be obtained by excavating one square kilometre to a depth of just two metres.

The essential elements in the construction scheme are the HLLV, an orbital transport system, a manned lunar base, an L-5 'construction shack', a couple of mass drivers, bulk carriers and a set of mass catchers. Present-day technology is adequate to develop and build each of these items. Given the funding and an intensive twenty- to thirty-year programme there seems to be no engineering reason why such a colony could not be built. If such a scheme were set in motion in the nineteen-nineties, the first fully fledged space colony could be functioning by about 2020. It could be earlier; you had better hurry if you want to get your name down!

Preliminary estimates indicate that the first colony could be established at a cost of about a hundred billion dollars (1975 values). We all know how little reliance can be placed on such estimates, but this gives some kind of figure to work on. It has been suggested that the colonies could repay the initial outlay in a number of ways, including the manufacture of specialized materials and components utilizing the advantages of the space environment, but a preliminary cost-benefit analysis* suggests that it should be possible to repay the entire capital outlay within 35 years of the commencement of the project (i.e., twelve years after completion of the first colony) if the efforts of the space community are directed to the fabrication of satellite solar power stations (SSPS), which would collect solar energy and beam it down in the form of microwaves to receiving stations on Earth.

*See, for example, the article by Mark M. Hopkins of Harvard University in the *Journal of the British Interplanetary Society*, volume 30, pp 289–300, 1977.

Plans are already well advanced for a system which could deliver between five and ten thousand megawatts (Plate 23), between a tenth and a fifth of the entire output of the present National Grid in the United Kingdom; and a smaller-scale experimental SSPS may be placed in synchronous orbit round the Earth before 1985. The advantages of beaming solar power to Earth are that it makes no demand on dwindling Earth resources; that the energy is 'clean', producing no toxic or unsightly wastes; and that there would be minimal effects on the environment because the amount of *waste* energy is likely to be far less than that released by fossil-fuel, or even nuclear, power stations. Barring fusion power, the only medium-term alternative to the SSPS— as fossil-fuel reserves dwindle—is nuclear fission with its hazardous and exceedingly long-lived radioactive products, the indefinite storage of which seems to be highly impracticable. At the very least, a rolling programme of SSPS construction would buy us the time to develop working clean fusion systems without creating problems for ourselves and future generations. In fact, one or two thousand SSPS would meet the entire present world energy requirement.

The colonies described above will be only the forerunners of much larger communities. O'Neill has envisaged a series of stages culminating in 'Island Three'—a colony of ten million people housed in a pair of cylinders each about 35 kilometres long by about 6 kilometres wide and laid out inside as three parallel valleys, separated by long windows through which sunlight would be reflected by means of long mirrors (Plate 37). Colonies on such a colossal scale could begin to make some contribution (assuming zero population growth on the Earth) to reducing the global population towards the optimum value, and within such communities there would be tremendous scope for a wide variety of leisure and cultural pursuits. Nevertheless, in a confined physical environment, a sense of purpose and of *contributing* to the community would be even more important than it is here on Earth; it would be hard to 'drop out' in such a colony.

It is too early to say whether or not the development of our space environment will proceed in this way; perhaps the exploitation of space will proceed initially without the construction of self-sufficient colonies. Although initially cheaper, the latter approach must be more expensive in the long run due to the greater costs of transporting men, supplies and materials from the Earth. Colonies make sense, and it may be that, if we survive as a species, only the details and timescale will be open to debate.

Manned Planetary Missions
Following after the remote robot probes and rovers it seems inevitable that Man will wish to explore the Solar System in person. In terms of the utilization of the easily accessible resources of the Solar System, expeditions to the asteroids will be a logical step, and we may well see

schemes to transport whole asteroids to the vicinity of Lagrangia, or—on the other hand—we may see colonists setting out to build a habitat inside a small asteroid and then transporting it slowly to whatever point in the Solar System happens to suit their convenience. But these are projects for the latter part of the next century.

Nearer our own time, we shall almost certainly see a lunar base established by the early nineteen-nineties, either as part of the space colonization project or as a separate scientific venture. Manned expeditions to the nearer planets may commence about the end of this century, with a Mars landing mission being the obvious first step. Although such a mission is not yet a budgeted item a fairly detailed NASA blueprint has existed for some time for a manned voyage involving a combined crew of twelve aboard two spacecraft powered by nuclear rockets. A Hohmann transfer orbit would give the vehicles a journey-time of about 270 days, at which stage they would enter orbit round Mars, remaining for about eighty days, during which time surface expeditions would be undertaken by means of lander vehicles. The return flight would utilize a higher-energy transfer orbit to obviate the need to wait more than a year before setting out; the entire mission would last nearly two years.

With new developments likely in propulsion technology, it may well be that the actual Mars expedition will be undertaken in rather different fashion, but it is unlikely to be more than a few decades away from today. By the end of the twenty-first century, we may expect manned expeditions to have encompassed the entire planetary system, just as we are anticipating this stage being reached by unmanned probes by the time the present century is out.

The second Mars expedition, early in the twenty-first century, will almost certainly lay the foundations of a manned base on that planet, and manned bases are likely to be established far and wide in the Solar System as the century progresses. Whether or not the planets will be *colonized* by large numbers is more problematical. With colonies in space it should be possible to optimize living conditions—in the material sense at least—and to exert complete control over the environment; with planets the situation is different. In the case of Mars, for example, colonists would have to contend with a thin unbreathable atmosphere, which nevertheless is capable of whipping up planet-wide dust storms, and extremes of temperature. As with the Moon, communities would be established within manmade structures, but nothing could be done about the reduced gravitational field which—if the Skylab results are extrapolated—may lead to physiological problems in the long term.

However, planets do have large surface areas and offer much scope for exploration. For those to whom the claustrophobia of the space colony is unacceptable, life on another planet might be a better alternative; or, again, those for whom a sheer physical challenge is important may find the satisfaction they seek from life in pushing back the frontiers of another world. To be confined to enclosed environmental structures and to be burdened with the necessity of wearing life-support systems whenever an expedition out of doors is contemplated would be inhibiting, but perhaps something could be done to change the entire environment of the planet to suit human requirements. With a nasty world like Venus, this would be a necessary prerequisite to any kind of colonization effort.

The process of changing the natural environment of a planet to something resembling conditions on Earth has come to be known as terraforming, and several ingenious schemes have already been suggested. Carl Sagan considered this possibility as far back as 1961 and, in his book *The Cosmic Connection* (1975), has put forward an outline plan for modifying the atmosphere of Venus.

Venus is a hostile world, shrouded in a dense carbon dioxide atmos-phere with clouds made up of droplets of sulphuric acid, and with a surface temperature of about 750 K. It is generally agreed that this high temperature is maintained by the 'greenhouse effect', whereby the Venusian atmosphere lets in short-wave radiation but prevents most of the heat escaping into space; the carbon dioxide content of the atmos-phere is the principal cause of this. Remove the carbon dioxide and the temperature and atmospheric pressure would both drop.

Sagan's suggestion is to inject into the upper atmosphere large quan-tities of a species of algae which would carry out photosynthesis in the vicinity of the clouds, converting carbon dioxide and water into organic compounds and oxygen. As the algae drifted down into the hotter lower regions of the atmosphere they would be fried, releasing carbon, carbon compounds and oxygen. The net result would be to maintain the overall water content of the atmosphere while converting carbon dioxide into carbon and oxygen. As the carbon dioxide content was reduced the temperature would decline, and as the oxygen oxidized the surface rocks (as it has on Mars) oxygen would be removed from circulation and the atmospheric pressure would fall. Provided the algae could reproduce faster than they were destroyed they could—in principle—convert the atmosphere to something more acceptable to us in a geo-logically short time, perhaps even within a few centuries.

I cannot believe it would prove so simple in practice, but such a scheme serves to remind us that planetary environments can be changed—and quite dramatically—in relatively short periods of time. We are doing it now with the Earth in a multitude of ways. By adding heat directly through industrial processes and domestic heating, and indirectly by the addition of carbon dioxide to the atmosphere by the burning of fossil fuels, we are raising the global temperature to a significant extent; recent estimates suggest that if we continue this process for a further fifty years the temperature in Antarctica may be raised by as much as 10C°. This would lead to the melting of sufficient ice to raise the mean sea level by about five metres, with consequences which are self-evident. By clearing woodlands we are depleting a source of atmospheric oxygen, and by liberally spraying everything in sight with assorted potions from aerosol spray cans we may even be depleting the atmospheric ozone layer which protects us from solar ultraviolet light. Most manmade effects on the environment have become apparent only in the last century; let no one doubt the feasibility of deliberately and drastically modifying a planetary atmosphere!

Schemes proposed for Mars include the heating of surface rocks to release oxygen and water vapour, the melting of the ice caps to achieve the same end, and the direction of cometary nuclei or material from the icy satellites of the giant planets onto its surface so as to raise the atmospheric pressure and allow liquid water to exist on the surface. The low surface gravities of the Moon and Mercury, together with the latter's proximity to the Sun, preclude the possibility of supplying them with a permanent atmosphere (atmospheric gases would rapidly leak away) and, although the larger satellites of Jupiter or Saturn may offer some prospect for terraforming, the technology required would be of a very high order compared to the straightforward construction of space colonies.

Terraforming may offer some prospects for the colonization of Mars or Venus, but this would offer no way around the zero population growth requirement. At today's growth rate both planets would be groaning at the seams within about fifty years. I am not greatly enthusiastic about terraforming: it would be more difficult than the construction of colonies, and it would take much longer; it offers no solution to the expanding human population. Humanity would have attained zero growth rate or perished long before any such scheme could reach fruition; the problem of gravity cannot be overcome. Above all, the precipitate conversion of planets to suit our short-term needs might be construed by later generations as the most wanton act of vandalism.

If space colonies prove to be psychologically unacceptable then terraforming may offer the means of colonizing part of the Solar System, if that is still deemed to be a worthwhile aim in these circumstances. It should not, in my opinion, receive a high priority otherwise. The establishment of colonies or bases *on* planets is quite another matter; it is highly desirable and, I am sure, will happen.

Possibly the future colonists who take the road to the stars will find terraforming an acceptable procedure to apply to planets in their target systems. If so, I hope that the interests of any indigenous life-forms will remain paramount, and that terraforming will not be undertaken if it leads to the destruction of an alien species. After all, if the Alpha Centaurians (carbon dioxide breathers!) should arrive on Earth and without so much as a 'by your leave' begin to break down chalk and limestone to restore the Earth's primitive carbon dioxide atmosphere, we should be more than a little upset. The responsibility for undertaking terraforming could only be assumed after prolonged debate and careful assessment of all possible objections. We could expect—and quite rightly—the emergence of conservationist groups dedicated to preventing the rape of the planets. On Earth at the moment we are concerned about the rapid loss of wilderness, the diminishing areas of unspoiled terrain which have escaped the more obvious of Man's ravages. If terraforming is not essential, and if we can accommodate ourselves with greater comfort and convenience in manmade space colonies, perhaps we should try to maintain the planets in their natural states.

The Outlook for Home Territory

In summary, the twentieth century will see unmanned probes pass the boundaries of the planetary system, and will see Man step firmly into the arena of space utilization with manned orbiting laboratories, an extensive network of applications satellites and the capability with the Shuttle and HLLV to assemble large structures in space. Space industries will move from the experimental to the developmental phase, and the Third Industrial Revolution will be under way. The lunar colony will be established, and the construction of manned space colonies may well have begun.

Next century should see manned expeditions traverse the entire Solar System and the establishment of manned bases on some of the planets, while space colonies may grow in scope and numbers. The Earth itself, its ecosystem saved from catastrophe by the shifting of industrial activity away from its surface and by the beginnings of a reduction in population, may well have become the most delightful and restful backwater to which the space colonists may flock for adventure holidays, risking the hazards of an uncontrolled biosphere to visit the beautiful but relatively savage world from which their ancestors stepped out into Man's larger environment.

But by this time, and probably before, Man's exploratory urge will have sent him on his first interstellar voyages, cautiously at first with unmanned probes, but later with manned starships or self-propelled colonies. The home territory of the Solar System will surely become the new cradle out of which humanity needs must climb.

6 Proxima and Beyond — the Problems

When Man sets out to cross the abyss, whether directly in manned starships or by proxy using automatic probes, the problems encountered will be many orders of magnitude greater than those related to the exploration of our cosmic back-yard, the Solar System. The fundamental problem is *distance*. Proxima Centauri, the nearest star, is 4.3 light years away, and such a distance is 270,000 times further than the Sun, nearly seven thousand times farther than Pluto, and well over one hundred million times farther than the Moon, the only other body on which Man has set foot to date.

Let us imagine a scale model of the Solar System in which we represent a distance of 100,000 kilometres by 1 metre, so that the Earth is represented by a small globe just over 12 centimetres in diameter; the Sun lies at a distance of 1500 metres; and Pluto is at a range of 60 kilometres. On this scale, Proxima Centauri would be located at about the *actual* distance of the Moon and we should need a spacecraft to get to that star even on our model. Our Galaxy's diameter is represented on the model by a distance greater than the actual distance of Pluto. Those who indulge in the sport of jogging can probably jog their way for 1500 metres in about 8 minutes, in which case they would be moving on the model at the scale speed of light. The most ardent jogger would balk at continuing at this pace for 60 kilometres (although a marathon specialist might cope) to cover the scale distance to Pluto, but to jog at a steady pace (without stopping) for 4.3 years to reach our model of Proxima Centauri would defeat even the most enthusiastic candidate for inclusion in *The Guinness Book of Records*.

Our first interstellar probe, Pioneer 10, which flew past Jupiter in December 1973 and was accelerated by the encounter in the manner described in Chapter 4, having exceeded the escape velocity of the Solar System is now settling down towards a steady speed of about 11.5km/s (i.e., 41,000km per hour) away from the Sun. At this speed it would take 110,000 years to reach Proxima Centauri—even if it were aimed in that direction, which it is not.

This would seem to pour cold water on the idea of interstellar probes. With the best will in the world, no agency is going to fund a mission whose results will take 100,000 years or more to return. With nuclear or ion rockets that are presently being developed we may reasonably hope to improve these times by a factor of ten or so, but that does not greatly help, since ten thousand or even one thousand years is still an inconceivably long time from the human point of view. To tread the road to the stars, other than by means of which we as yet have no conception ('space warps', 'hyperspace' and the like), seems to require us to choose between one of two options—to plan on taking hundreds or thousands of years, in which case only the remote descendents of the original crew will reach the target; or to devise means of propulsion to attain a sizeable fraction of the speed of light. The former 'solution', the *interstellar ark*, is one which we shall return to later, but for most people, I am sure, the latter solution must be the only acceptable one; who today would embark on a journey, knowing that he would never arrive, let alone return home?

Perhaps some major breakthrough in medical science or genetic engineering will increase the average human longevity to such an extent that journeys of a hundred or even a thousand years would fall within a human lifespan. After all, the problem is not so much the average duration of the journey as its length in relation to a human lifetime; any journey that occupies more than a few per cent of an individual's life is, for the moment, quite unacceptable. Such developments may come, but there is little evidence to suggest that we should be optimistic about the possibility in the foreseeable future.

It seems to me that the only acceptable way forward is to devise the means to travel at speeds not less than one-tenth of the speed of light (we shall denote this figure by $0.1c$, where c is the speed of light, about 300,000km/s). At $0.1c$, Proxima Centauri would be reached in 43 years, neglecting acceleration and deceleration. The closer we can get to the speed of light the better, but even so, so far as the Earth-bound mission controllers are concerned, interstellar journeys will be long-term affairs, and journeys beyond a few tens of light years will mean that no single controller can see the mission through from beginning to end. For the onboard crew, however, due to effects which arise from the Theory of Relativity and which are described later, the situation may not be nearly so bad. Faster-than-light travel does not appear to be possible within the framework of known physical laws.

Is it not rather academic to be discussing the possibility of travelling close to the speed of light when the best we can attain at present (with Pioneer 10, for example) is less than one twenty-thousandth of this? I don't believe this to be so. On 17 December, 1903, at Kitty Hawk in North Carolina, Orville Wright made the first sustained power-driven flight at a speed of about 55 kilometres per hour (airspeed), while just over 55 years later, on 2 January, 1959, the Soviet spaceprobe Luna 1 achieved the terrestrial escape velocity of 11km/s (40,000kph). In just over half a century an increase in speed of a factor of nearly a thousand was attained, and we may expect a further factor of ten increase by the end of this century with, for example, the SEPS. A further factor of one thousand would give a velocity of about one-third of the speed of light, and it does not seem impossible that the means to achieve this velocity could be available within a century, or even sooner.

In the next chapter we shall examine some of the possible techniques of interstellar propulsion which are currently being discussed but, for the moment, let us look at the question of the 'light barrier'. Why should the speed of light represent an unattainable velocity, a barrier which no material object (nor any known form of signal) may surmount? The barrier arises in the Special Theory of Relativity as presented by Albert Einstein in 1905, a theory which related the observations of events made by so-called *inertial* observers, observers moving at constant velocity relative to each other. What, one might ask, was so wonderful about that; surely it is a matter of elementary common sense to relate the observations of the same events made by observers moving at uniform speed? What Einstein's theory showed was that what passes for 'common sense' is not necessarily the best guide to what happens in the Universe, and that what is 'obvious' or 'self-evident' may be very far from the truth.

According to the classical, or Newtonian, view of the world, space and time were 'absolute'; that is to say that time was something which flowed by at a uniform rate wherever you might be and however you might travel, and that there existed some kind of 'absolute space', a sort of fundamental framework to the Universe against which, in principle, you could measure your absolute speed. Obviously it is difficult to tell how fast you, the reader, are moving; if you are sitting at home you will be moving because of the rotation of the Earth on its axis, the revolution of the Earth around the Sun, the motion of the Sun in the Galaxy, the motion of the Galaxy relative to the other galaxies, etc.

But, according to classical theory, it should be possible to measure the Earth's velocity through space or through the 'aether'—an imaginary fluid with which space was thought to be filled in order that lightwaves could be propagated through the void. All attempts to measure the velocity of the Earth through the aether failed, and experiments (such as the Michelson–Morley experiments of 1881 and 1887) indicated that the velocity of light was constant no matter what the speed of the observer or the source of light. This observation appears absurd and in flat contradiction to everyday experience. If two cars are approaching each other, each travelling at 80kph, and they collide, then the relative velocity of the collision must be 160kph—a simple case of $1 + 1 = 2$. Common sense tells us that, if a source of light is travelling towards us at, say, half the speed of light ($0.5c$) and it is emitting light (velocity $= c$), then the light which arrives should be approaching at one and one half times the velocity of light (i.e., $1.5c$). The observations indicate quite clearly that this is not so; the measured velocity is just c; $1 + 0.5 = 1$, apparently.

The result seems absurd, and when a scientific experiment produces a result in flat contradiction to accepted theory, the natural and sensible reaction is to doubt the validity of the experimental results. But when an experiment is checked, repeated, refined and repeated in different ways, and still comes up with the same result, then it is time to change the theory. The experiments indicate quite clearly that the common-sense approach is inadequate.

In setting up his Special Theory of Relativity, Einstein accepted the constancy of the speed of light as one of his two basic postulates (i.e., the measured velocity of light is constant and independent of the speed of source or observer). The other postulate was that 'all inertial frames are totally equivalent for the performance of all physical experiments'; i.e., that the velocity at which a laboratory is moving (provided it is constant) has no effect on the results of experiments carried out within

15
An artist's impression of the Space Shuttle in flight. The vehicle is launched with all engines burning, a configuration known as parallel burn. The two solid-fuel rockets, shown here attached to the larger liquid propellant tank, are jettisoned at an altitude of approximately 3000m; they drop by parachute into the sea and so can be recovered and reused. The orbiter discards the liquid propellant tank just before insertion into Earth orbit. (*NASA.*)

16
A cut-away view of the Space Shuttle. (*NASA.*)

17
An artist's impression of a Space Shuttle orbiter approaching a landing field following a spell in space; the orbiter will land on a conventional runway similar to that used by the jet aircraft of today. (*NASA.*)

18
With manipulator arms extended, the Space Shuttle orbiter prepares to take on board an orbiting satellite. A key feature of the Shuttle system will be its capability of retrieving payloads in orbit for repair or maintenance in space or for return to Earth. (*NASA.*)

19
An artist's impression of Spacelab in the payload bay of the Space Shuttle. Scientists and payload specialists are working in the pressurized forward module which is connected to the Shuttle orbiter airlock by a crew access tunnel. At the aft end are two pallet sections on which are mounted various scientific instruments, including telescopes, sensors and antennae, all exposed to the vacuum of space. The equipment on the pallet is operated remotely from the pressurized module. (*NASA.*)

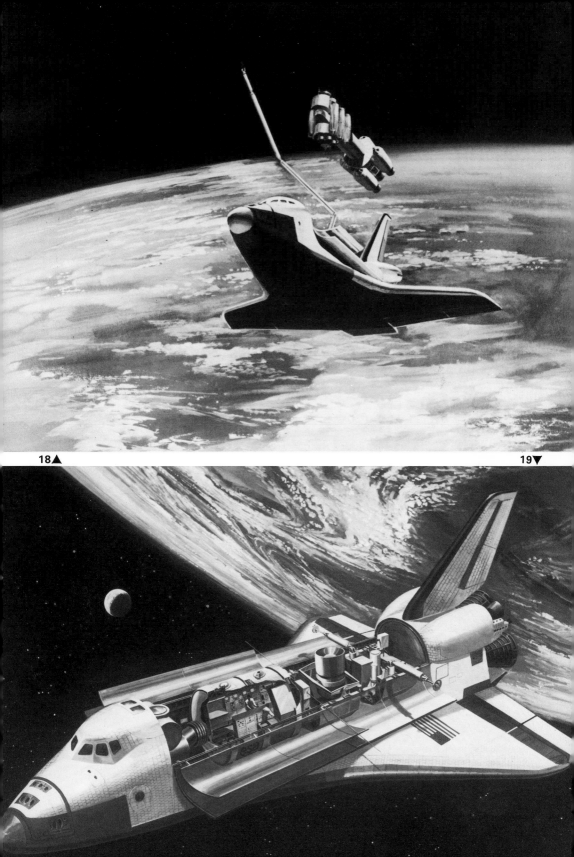

it. As a result, the idea of absolute space, and of the aether, was abandoned, and it was appreciated that the only measurements which can be made are *relative* ones, not absolute ones.

From this theory there stems a number of consequences of vital significance to interstellar travel, and these are itemized below.

Length Contraction

As the Irish physicist G. Fitzgerald had hinted earlier, the length of a moving object is affected by its motion. A spacecraft moving past an observer (whom we shall describe as 'stationary', although accepting the notion that nothing in the Universe can claim truly to be 'fixed') at high speed will appear to the fixed observer to have shrunk in length by an amount which depends on the speed of the spacecraft. The closer the spacecraft approaches to the speed of light the more pronounced this effect becomes until—if it were possible for it to move precisely at the speed of light—it would have zero length. Values of length contraction for different velocities (together with the formula for calculating the effect) are given in Table 4.

It is important to note, however, that the inhabitants of the spacecraft will not be aware of this effect. To them everything is normal, for the 'contraction' will have affected everything in the spacecraft in equal proportion.

20
Artist's impression of two of the necessary steps leading to the development of a fabrication and assembly facility in space. The first flight demonstration would involve producing several structural members by an automated beam fabrication module in the Space Shuttle cargo bay. The second flight would demonstrate beam fabrication plus limited erection and assembly. These are two of the stages towards construction of a large structure, such as a solar power station to provide electrical power in space. This demonstration could begin in 1980 and culminate in a major demonstration during 1983–4. (*NASA.*)

21
Artist's impression of two of the proposed Teleoperator Space Spiders (one at the background left of the disk, the other at the foreground right of the disk) building a structure for a solar power satellite onto a Space Shuttle external tank. The Spiders, self-contained systems with rolls or spools of coded and prestamped building materials, would be capable of forming and assembling a structure in one integrated operation. With such a structure completed the external tank could become a control centre for space operations, a crew habitat for Shuttle astronauts and a focal point for various space missions. (*NASA.*)

TABLE 4

RELATIVISTIC EFFECTS

Velocity (v) as a fraction of the speed of light (c) $\dfrac{v}{c}$	Length (l) of a moving object as a fraction of its rest-length (l_0) (length contraction)	Mass (m) of a moving object compared to its rest-mass (m_0) (mass increase)	Length of a time interval measured on the moving clock compared to the time interval measured on Earth (time dilation)
0	1·000	1·000	1·000
0·1	0·995	1·005	0·995
0·3	0·954	1·048	0·954
0·5	0·867	1·155	0·867
0·6	0·800	1·250	0·800
0·7	0·714	1·400	0·714
0·8	0·600	1·667	0·600
0·9	0·436	2·294	0·436
0·95	0·312	3·203	0·312
0·99	0·141	7·089	0·141
0·999	0·045	22·366	0·045

The terms *rest-length* (l_0) and *rest-mass* (m_0) refer to the length and mass which the moving object would have if it were stationary relative to the Earth. The relationship between length (l) and l_0, mass (m) and m_0, and a time interval (Δt) measured on the moving clock and a time interval measured on Earth (Δt_0) is given by the following:

$$ l = l_0 \sqrt{1 - \frac{v^2}{c^2}}, \quad m = \frac{m_0}{\sqrt{1 - \frac{v^2}{c^2}}}, \quad \text{and} \quad \Delta t = \Delta t_0 \sqrt{1 - \frac{v^2}{c^2}}. $$

Time Dilation

The rate at which time passes on a fast-moving spacecraft is slower than the rate at which time passes as measured by a 'stationary' observer in his own frame of reference. If the Earth-based observer could see the clocks on the fast-moving spacecraft he would come to the conclusion that they were running slow compared to his own clock. Again, this effect (see Table 4) becomes rapidly more obvious as speeds approach that of light until, if it were possible for a spacecraft to move at the speed of light, time inside it would stand still compared to the stationary observer's time.

This affects everything. Not just mechanical clocks but also atomic processes, all phenomena within the spacecraft and, above all, the biological clocks of the inhabitants of this hypothetical spacecraft, will be affected in equal proportions. Thus, although everything seems perfectly normal to the astronauts, so far as the Earthbound observers are concerned the astronauts are ageing more slowly than their terrestrial compatriots. If one member of a pair of twins were to take a long journey at close to the speed of light, he would return to find his stay-at-home twin much older than himself. Consider the hypothetical example of Fred and Dave Lux: Dave prefers the easy life at home while Fred, on their thirtieth birthday, sets out on a voyage to a star 14 light years away in a spacecraft which travels at 99.9% of the speed of light. Neglecting acceleration, deceleration and turnaround time, he flies directly to the star and immediately sets out on the return trip, arriving back after 28 Earth years to find Dave now aged 58. For Fred only four years will have elapsed and he will be 34 years of age. The result may appear ridiculous, but the time-dilation effect has been confirmed in a number of different experiments, and there seems to be no doubt that what we have described is just what would happen in a real situation.

In the absence of some hitherto unsuspected loophole in the 'light barrier', the time-dilation effect offers the only way in which an individual astronaut can make a long interstellar journey and reach his target, or return to Earth, within his natural lifetime. By moving fast enough, *any* journey is possible. The penalty which must be paid—for in nature one cannot truly get something for nothing—is that an astronaut cannot return to his own time on Earth. On a round trip of a few decades of Earth-time, he might expect to return to some familiar, if older, faces, but on journeys over hundreds or thousands of light years (which might occupy only a few years or decades on the starship) there would be no knowing what, if anything, or anybody, would await the returning explorer. Perhaps the long-range relativistic traveller will become an interstellar nomad, choosing never to return to his native home because, to him, it has ceased to exist.

Mass Increase

An unfortunate effect from our point of view is that the mass of a moving body is greater than its rest mass (i.e., the mass which it would have if it were stationary in the observer's frame of reference). The closer the spacecraft approaches to the speed of light the greater its mass becomes until, if it could reach the speed of light, its mass would become infinite. This is serious from the point of view of interstellar travel for it means that the amount of energy required to accelerate further a spacecraft already moving near to the speed of light is very high, and it is impossible to supply sufficient energy to accelerate it all the way to that magic velocity. Given the resources, the speed of light may be approached as closely as you wish, but it can never be attained. The speed of light is truly a barrier in theory as well as practice. We may utilize time dilation to enable astronauts to travel as far as they wish, but we cannot bring them back to their own time again.

The Equivalence of Mass and Energy

Einstein showed that matter and energy are interchangeable, and that if a certain amount of mass (M) were converted into energy, the amount of energy released (E) was given by $E = Mc^2$, where c denoted the speed of light. Since the speed of light is a large number and the speed of light squared a very very much larger number, it follows that a great deal of energy may be obtained from the destruction of a quite small amount of matter. It was this relationship which allowed us to understand how the Sun and stars were shining, and how to harness the energy of the atom, and which may provide us with the means to reach the stars. On the dark side, it has also led to the development of the atomic bomb which, when it became a fearful reality, led Einstein to remark: 'If only I had known, I should have become a watchmaker.'

Even the Sun is relatively inefficient in applying this process, for only 0.7% of the matter involved in nuclear reactions is actually converted into energy. We are still some way from duplicating even this degree of efficiency here on Earth, but there is hope that we may do so before long. $E = Mc^2$ provides the clue to obtaining colossal amounts of energy, and must be a key relationship in taking us along the road to the stars.

The 1g Spacecraft

Standing on the surface of the Earth, we are aware of our weight, and the longer we stand still the more aware we become! The feeling of weight arises because, by standing on a solid surface, we are resisting the gravitational force exerted on the mass contained in our bodies, which is attracting us towards the centre of the Earth. As we saw earlier a body falling freely under gravity experiences no sensation of weight and accelerates towards the attracting mass at a rate equal to the 'acceleration due to gravity' which, in the vicinity of the Earth, is about 9.8 metres per second per second, denoted by '1g'. To someone inside a closed box, the effects of a gravitational field are indistinguishable from the effects of acceleration; an astronaut in the depths of space whose spaceship is accelerating at a rate of 1g will feel a sensation of weight exactly equal to his normal weight on Earth.

It seems likely that prolonged periods of weightlessness will have physiological effects, some of which may be irreversible, that may make it hazardous for astronauts to return to Earth (for example, the loss of bone calcium mentioned earlier would lead to brittle bones), and for that reason alone it may be necessary to generate artificial gravity in long-range spaceships. One way to do this has already been discussed in the previous chapter; i.e., spinning the space vehicle. For a starship, however, probably the ideal solution would be to have the spacecraft under continuous powered flight, accelerating at a constant rate of 1g so that the astronauts feel their normal weight. Present-day rockets can achieve this sort of acceleration for only very limited periods (measured in minutes), while the proposed SEPS system can sustain only tiny acceleration rates. In the next chapter we shall examine whether or not the goal of 1g flight can be attained.

For the moment, let us assume that it can, and examine the consequences of maintaining 1g acceleration indefinitely. We will see some startling results which seem to render all journeys possible. In everyday experience, the relationship between acceleration, time and velocity is straightforward: the velocity reached from a standing start after a given time at a constant acceleration is just equal to the acceleration multiplied by the time. The results of acceleration at $1g$ are such that after about one month a speed of about 90 million kph would be attained, and the speed of light would be reached in just under one year.

Before the speed got close to that of light, relativistic effects would come into play to modify the relationship between acceleration, time and distance (for the spacecraft cannot attain the speed of light) in such a way that the measurements of these quantities made by the crew would be quite different from those made by Earth-bound observers. The time dilation effect becomes greater the longer the spacecraft has been travelling, with the results indicated in Table 5. For example, in 5 years of ship time the spacecraft can reach out to about 30 parsecs and 100 years of Earth time will have elapsed; 12 years of ship time would suffice to cross the entire Galaxy, during which period 100,000 years would have passed on Earth! Of course, these figures apply to constant acceleration. On a landing mission, the spacecraft would be accelerated at $1g$ for half the distance, and decelerated for the other half so as to achieve a soft landing. This would increase the mission time. For example, it would take more than 10 years of ship time to reach, and land upon, a target planet at a range of 60 parsecs, and a round trip involving a return to Earth would occupy rather more than 20 years (compared to just over 400 years Earth-time). Nevertheless, the figures are impressive, for even a round trip to the Andromeda Galaxy could be accomplished in about 60 years.

TABLE 5

INTERSTELLAR VOYAGES AT CONSTANT $1g$ ACCELERATION

The values of elapsed ship time and range attained apply to continuous acceleration, and do not allow for deceleration at the target (see text).

Earth time (years)	Ship time (years)	Range attained	
		(light years)	(parsecs)
1	0·9	0·4	0·2
10	3·0	9	2·75
100	5·2	99	30·6
1000	7·5	999	306
10000	9·7	9999	3067
100000	12	10^5	3×10^4
1000000	14	10^6	3×10^5
10^{10}	24	10^{10}	3×10^9

There are massive problems in the way of achieving this constant acceleration, and if it is ever attained it will probably have to be with a spacecraft that picks up its fuel on the way; no vehicle which has to use up most of its propulsive energy simply to accelerate a huge mass of fuel can hope to sustain $1g$ for very long. Perhaps it may never be achieved, but who can state dogmatically that it is *impossible*? $1g$ acceleration seems ideally suited to human interstellar exploration for the twin reasons that it gives us a normal feeling of weight and that it allows any interstellar journey to be accomplished within the span of a single human lifetime.

Heating, Drag and Erosion

We tend, naturally, to imagine that in travelling through space we will be free of the problems which we experience in the Earth's atmosphere, such as heating due to atmospheric friction, drag (the resistance experienced by a moving body in air) and erosion (the wearing away due to impacts of dust particles). At conventional speeds, this is certainly so (although erosion does occur to satellites due to the impact of micrometeorites), for space is the closest approach to a complete vacuum that we can find. But space is not entirely empty, interstellar space being filled with a tenuous mixture of gas and dust as described earlier.

It has been calculated that for a vehicle travelling at one eighth of the speed of light, the delay due to drag over a fifty-year journey would amount to only a few *minutes* (see Project Daedalus in the next chapter).

Due to the effect of length contraction, as the speed of the spacecraft increases, the apparent density of interstellar matter increases (from the spacecraft's point of view, distances are reduced, and the interstellar material is compressed into a smaller volume) and this gas, moving relative to the spacecraft at near the speed of light, is capable of producing a tremendous amount of heating. Various methods of reducing this effect have been discussed, including the use of heat pipes to take excess heat from the forebody to radiators which would dissipate it into space, and the carrying of a device to ionize the interstellar gas ahead of the starship, the ionized gas being deflected by some kind of magnetic shield.

The heating, and the less significant effect of drag, can both be reduced by streamlining, and the surprising outcome is that high-speed interstellar spacecraft may have to be streamlined just as effectively as are high-speed aircraft. Slow-moving interplanetary spacecraft (such as the well-known Apollo Lunar Module) require no aerodynamic properties; relativistic starships will need the most careful design.

A starship travelling over many parsecs at speeds approaching that of light will be struck by a stream of interstellar dust grains, charged atomic particles, hydrogen atoms and so on, all of which will be moving relative to the ship at such high speeds that their masses (and hence their energies) will be enhanced by the relativistic mass effect. These will erode the surface of the spacecraft and, although there is considerable debate about the magnitude of the effect, it seems likely that shielding will be required, possibly in the form of a 'precursor shield' which travels ahead of the actual starship, absorbing these impacts.

It would appear that charging around the Universe is not the straightforward business it might be thought to be.

The Economics of Building Starships

Obviously starships will be expensive. At this stage it would be impossible to produce a realistic costing for such a project, even for one as closely specified as Project Daedalus. We seem to have a singular lack of success in the advance costing of major technological projects, a good working rule being to take the estimate and multiply by a factor of between two and ten.

We can at least look at some general principles. The cost of the Apollo project has been estimated at around 25 billion US dollars; this figure includes the development of hardware, ground control, launch and tracking facilities. Probably the best indicator of the economic resources of a nation is its gross national product (GNP) which is basically a measure of the country's output. In 1971, while Apollo flights were still being carried out, the GNP of the USA was about 1000 billion dollars, and the *total cost* of Apollo amounted to about 2.5% of this. However, since Apollo was spread out over more than a decade, the spending on the project in any one year was less than 1% of the GNP. Let us assume that a wealthy nation can afford to spend 1% of its GNP on a major technological venture—if it so desires.

The cost of developing and building the first starship capable of undertaking a mission within a reasonable period of time is unlikely to be less than the cost of building the first space colony for 10,000 people, and the estimate for that project is about 100 billion dollars (at 1975 values). This is about 10% of the US GNP, and clearly such an undertaking is just out of the question at the present time. Since the USA is by far the richest nation in the world, even a global commitment, today, would be unacceptable.

However, the factor which can change all this is economic growth, the growth of a nation's output. No one seems to be quite sure what causes economic growth—technological advances, investment, education all seem to be factors—but, whatever its causes, economic growth exists and provides us with a measure of the increase of a nation's (or of the world's) wealth. Even after making allowance for inflation, there has been considerable net economic growth among the industrialized nations since the coming of the Industrial Revolution. In the United States economic growth rates of around 5% have occurred, and even a pessimistic estimate anticipates a continuing net growth rate of 2% per annum. Ignoring the possibility of economic collapse and the general collapse of civilized society, which latter remains a real possibility, let us see the outcome of such a growth rate. GNP will double every 35 years, increasing by a factor of 10 before the year 2090. Table 6 shows the results of economic growth rates of 2% and 5%, clearly demonstrating the power of compound interest—at the higher rate, GNP will be increased by a factor of 10 by about 2020, and by a factor of 1000 in only 142 years! Even the lower rate leads to a hundred-fold increase in wealth in 233 years.

If the first starship cost ten or even a hundred times the outlay on the Apollo programme, that should present no insuperable burden after a century or so has elapsed from today—*provided that economic growth is sustained*. If we overcome our immediate crises, and really move out

TABLE 6

GROWTH IN GNP WITH TIME

GNP (Gross National Product) expressed in units of 1000 billion US dollars (i.e., starting at a GNP of 10^{12}).

Date	2% annual growth	5% annual growth
1971	1	1
1990	1·46	2·53
2010	2·16	6·71
2030	3·22	17·8
2050	4·78	47·3
2070	7·10	126·0
2090	10·6	333
2113		1000
2204	100	

into the space environment as discussed in Chapter 5, then the new industrial revolution may well bring about a high economic growth rate (without the undesirable side effects on the Earth's environment) which could bring forward the construction of the first starships. It may also be that the cost involved in the construction will be less than the figures I have suggested, particularly if the necessary technology (e.g., fusion power) is developed in other contexts.

· The quite reasonable conclusion is that by the middle of the next century we may well be able to afford to build starships.

Target Stars

Given the starships, where should we send them? Particularly in the early days of interstellar exploration, when the cost in time and money of each spacecraft will be prodigious, there must be rigorous criteria for the selection of target stars. These will include:

(*a*) distance: clearly the nearer the star happens to be the better;
(*b*) the astrophysical interest of the star itself;
(*c*) the existence of a planetary system around the star; and
(*d*) the likelihood of one of these planets being suitable for life and/or already supporting life.

Factors (*a*) and (*b*) are already quite well known for nearby stars and we may expect knowledge of factor (*c*) to improve over the next few decades once space-borne telescopes and improved astronomical techniques come into play. At present, the only means of detecting planetary companions of stars is to look for the very slight 'wobbling' introduced to their motion through space by the gravitational attraction of these planets on their parent stars. Such observations are extremely difficult to make and must be spread out over long periods of time. Nevertheless, there are several nearby stars which, according to the observations, have planetary companions, and the best-attested example is Barnard's Star which—according to Peter Van de Kamp—has two planets comparable with Jupiter revolving round it. The periods of these planets are estimated to be about the same as those of our own giants, Jupiter and Saturn, in their orbits round the Sun and presumably, if there are massive planets like these, there are probably smaller planets like the Earth, too, although these would not be sufficiently massive to exert a measurable effect on the star.

As we saw earlier, factor (d) is difficult to estimate, but in view of the length of time it has taken for intelligent life to develop here on Earth then we ought at least to be looking for long-lived stars, and here we have some information to go on. The very hot and highly luminous stars (of spectral types O, B and A) may be eliminated on this score, and in any case they emit so strongly in the ultraviolet part of the spectrum that life as we know it might not be able to survive. Various writers have tended to eliminate binary and multiple stars on the grounds that—even if they have planets—the planetary orbits may not be sufficiently stable (this assumption has been shown to be invalid in the cases where either the two stars are very close together and the planet far away, or the planet is close to one star and the other star is far away).

The cool M-type main sequence stars are extremely long-lived, but planets will have to be very close in if they are to be warm enough for our kind of life. If we eliminate these as well (although many of them may have life-bearing planets) we are left with looking at F-, G- and K-type stars, and if we restrict the list still further by looking only at stars in the range Go (temperature 6000 K) to K5 (4000 K)—i.e., those which are closely similar to the Sun—then within a range of about six parsecs we are left with only six suitable stars (see Table 7) which are not members of binary or multiple systems.

A star of considerable interest, although it does not meet the fourth

TABLE 7

SUITABLE TARGET STARS WITHIN 20 LIGHT YEARS			
Star or stellar system	Distance (light years)	(parsecs)	Remarks
(a) Single stars (spectral type G0 to K5)			
Tau Ceti	11·8	3·62	spectral type G8, cooler than Sun
Sigma Draconis	18·5	5·67	K0
82 Eridani	20	6·13	G5, similar to Sun
Epsilon Eridani	10·8	3·31	K2, cooler than Sun and about one third of Sun's luminosity
Epsilon Indi	11·2	3·44	K5, about one tenth Sun's luminosity
Sigma Pavonis	18·5	5·67	G6, practically same luminosity as Sun
(b) binary or multiple systems of particular interest			
Procyon	11·3	3·47	F5 plus white dwarf companion
Sirius	8·7	2·67	A0, plus white dwarf companion
Alpha Centauri	4·3	1·32	Triple (G0, K5, M5), the M5 star being the nearest star to the Solar System (Proxima Centauri)
61 Cygni	11·0	3·37	K5, K7, suspected planets
(c) Other stars			
Barnard's star	6·0	1·84	M5, dull red star having at least two planets
Lalande 21185	8·2	2·52	M2, may have at least one planet

criterion and may not meet the third, is Sirius, the brightest star in the sky—a star of type A1, 26 times the Sun's luminosity and lying at a distance of 2.64 parsecs. It is of particular astrophysical curiosity in having a white dwarf companion; and, since there has been some speculation as to whether or not it has planets, and whether or not we may have been visited in the past by intelligent beings from this system,* at least we ought to return the visit.

No doubt, before the first interstellar probe is launched there will be much debate among the scientific community on these and other factors, and upon aims—will the primary goal be astrophysics or the quest for life? The British Interplanetary Society's Daedalus study selected Barnard's Star as the target, as being one of the nearest (1.84 parsecs distant) and as being the one where the evidence for planets is strongest. Future mission planners may come to quite different conclusions.

Interstellar Navigation

In order to reach the stars at all, navigation and guidance will have to be of the highest order. In its simplest terms the navigational problem is two-fold—firstly to know where you are at any instant, and secondly to know your speed, course and estimated time of arrival. How can this problem be solved?

Consider, for a moment, the problem of a small-craft navigator in confined waters. He can obtain a 'fix' on his position by taking the bearings of two fixed objects (the lighthouse and the church tower in fig. 19) and plotting these on a chart; where the two position lines meet, there is the location of his vessel. Of course, there are always observational errors to contend with, and a more reliable position may be obtained by taking bearings on three fixed points. The three position lines will not in general cross at a single point but will give rise to the shaded triangle, known as a 'cocked hat'; the vessel lies somewhere within that triangle, but the navigator cannot say for sure exactly where. The size of the cocked hat is a measure of the errors in his observations.

In principle the interstellar navigator can determine his position in three dimensions (the coastal navigator is interested in only two dimensions, since he is confined to the surface of the sea) by an analogous process whereby he takes bearings of at least three suitably positioned stars; this gives him his position *in space* relative to these stars (fig. 20). The first essential is a three-dimensional star chart with precisely determined star positions (i.e., distance as well as direction known for each reference star), together with supporting data which will allow the navigator to identify stars without confusion, such data taking the form of, for example, precise details of the spectrum of each star. The chart and the data would be stored in the ubiquitous onboard computer.

*See *The Sirius Mystery*, by Robert Temple, Sidgwick and Jackson, 1976.

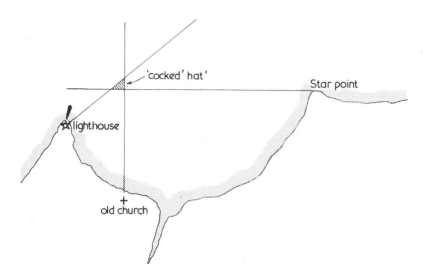

Fig 19 **Coastal navigation—position determination.** The navigator takes bearings on the lighthouse, church tower and headland ('Star Point') which show that he is somewhere inside the shaded triangle (or 'cocked hat'), the size of the triangle giving an indication of the errors of observation. A position obtained from only two bearings is generally less reliable.

Assuming the navigator (human or robot) can identify the appropriate stars then he can determine his position with precision only if the spatial positions of these stars are determined to accuracies typically a hundred times better than those so far attained. Even among the nearest stars, inaccuracies of a few per cent are common and fewer than 1000 stars have spatial positions known to even 10% accuracy. The precise determination of stellar positions will be an essential prerequisite task. The exploration of the Solar System will lead to much higher accuracies by allowing much longer baselines to be used for parallax measurements; for example, by establishing a number of telescopes spaced out round the orbit of Pluto an immediate improvement of a factor of 40 will be obtained (since Pluto is forty times further from the Sun than is the Earth). Long-range probes to the outer fringes of the Solar System (into the 'comet cloud') will provide even longer baselines, and the first interstellar missions would supply parallax data of ever improving accuracy as the distance between Earth and spacecraft increased. A baseline of 3 parsecs would give a precision 600,000 times better than we can now achieve, allowing us to make measurements of any star in the Galaxy to an accuracy far better than our current knowledge even of Proxima Centauri.

In principle the starship may be kept precisely on course by keeping the Sun dead astern and the target directly ahead, this being equivalent to the nautical procedure of sailing a transit line between two fixed marks. The speed and progress may be determined by making observa-

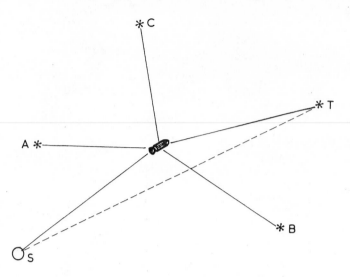

Fig 20 **Interstellar navigation—position determination.** By combining the bearings of three reference stars (of precisely known position) with bearings on the Sun and the target star, a good 'fix' may be obtained. The accuracy of the position will be improved by using more reference stars.

tions at intervals of the same set of stars and noting how their positions change—this gives a positional 'fix' at regular intervals, allowing speed and course made good to be evaluated. This is a slow and tedious method which may not give sufficient warning of velocity inaccuracies or of course changes which may be required.

With interplanetary spacecraft at present, the velocity is readily obtained by means of the Doppler effect, the change in wavelength or frequency of a signal due to the relative motion of the source and the observer. The onboard navigator in a starship can determine his velocity by measuring the frequency change in a signal of known frequency beamed from the Earth. He can also measure the Doppler effect in the spectra of light from stars (including the target) and confirm his velocity in this way even if he can't see the Earth (which is possible if his motors are firing or otherwise obscuring his view aft).

The principles of interstellar navigation are not greatly different from those employed in navigating a small boat within sight of known landmarks provided that the speed of the spacecraft is small compared to the velocity of light. At higher speeds two important effects will become increasingly obvious, the Doppler effect and aberration. The aberration of starlight is a phenomenon discovered in 1729 by James Bradley, later to become Astronomer Royal, and is the small displacement in the position of a star due to the combination of the finite velocity of light and the speed of the Earth in its orbit around the Sun. There is an everyday analogy here in the appearance of falling rain. Suppose that you are standing in a shower of rain with the raindrops falling vertically downwards. If you begin to run, the raindrops appear to be falling at an angle, and the faster you run, the more horizontal the direction of the raindrops becomes. Even if—due to the wind—the raindrops are falling at an angle *from behind*, if you could run fast enough they would seem to be falling at an angle from ahead (fig. 21).

The same sort of thing happens to light. The speed of the Earth in its orbit is about 30km/s, very small compared to the speed of light (300,000km/s), but it is nevertheless sufficient to produce a change in the apparent position of a star many times greater than its value of parallax. The higher the velocity, the more obvious the effect becomes, and as speeds approach that of light the effect becomes both obvious and very curious. The effect of aberration on a single star (fig. 22) is to make its apparent position lie closer to the direction in which the space-craft is heading, so, as speed builds up, the stars appear to be more strongly concentrated ahead while being spread out behind. Even stars which initially lie behind the starship would apparently be dragged forward so as to seem to lie in the forward hemisphere. If the stars were uniformly distributed around the sky when the spacecraft was stationary, at two-thirds of the speed of light something like 80% of the stars would seem to lie ahead, and at 90% of light speed all but 7% would have been dragged into the forward field of view. The hapless astro-nauts might well feel they were moving backwards!

Although the effects might be confusing, they should pose no real problems to the interstellar navigator and would, in fact, provide another means of confirming his velocity. The Doppler effect may have a more profound influence. As we have already mentioned, if a source of light of a particular wavelength is approaching us, the light waves are 'squashed up' and arrive with a shorter wavelength than the emitted wavelength; i.e., they are said to be 'blue-shifted'. Light waves from a receding source are 'stretched out' and arrive with a longer wavelength; these are said to be red-shifted. Stars lying ahead of the ship will be blue-shifted, since they will seem to be approaching the

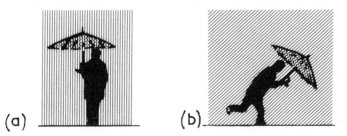

Fig 21 **Raindrops and observers.** The rain is falling vertically. In (*a*) the observer is stationary and notes that the rain *is* falling straight down, while in (*b*) the observer is running to the right with the result that the rain appears to be falling at an angle to the vertical as if it were coming from *ahead* of the observer.

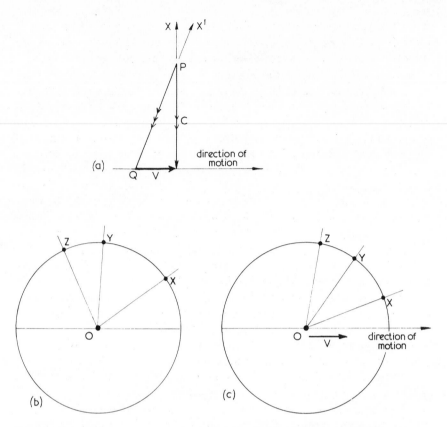

Fig 22 **The aberration of starlight.** In (*a*) the true direction of the star is X,
and arrow *c* represents the velocity of a ray of light coming from X; *v*
represents the velocity of the observer. As a result of the combination of
these two motions, the ray of light appears to approach the observer along
the line PQ; i.e., the star appears to lie in the direction of X'.

In (*b*) an observer aboard a stationary starship observes the positions
of three stars, X, Y and Z. In (*c*) the starship is moving with a velocity *v*
which is a large fraction of the speed of light and the apparent positions of
the stars are shifted as shown; even the star Z, which originally lay behind
the starship, appears to be ahead of it.

navigator, while stars behind will be red-shifted. The Sun, a typical
G-type star, emits most strongly at a wavelength of about 550 nano-
metres (near the middle of the visible spectrum) and if viewed from a
distance would appear as a yellow star for that reason. The human eye
responds to light in the wavelength range from about 400nm (violet) to
about 700nm (deep red); *if* a G-type star emitted light *only* at that one
wavelength (550nm), and if the Doppler effect were sufficient to shift
the wavelength of the incoming light beyond either of those limits, then
the star would disappear from view. If such a star were dead ahead or
dead astern *it would disappear from view* at a speed of about 30% of the
speed of light!

If all stars emitted only at this one wavelength the effect as the velocity increased would be dramatic indeed. More and more of the stars ahead and astern would disappear to give the appearance of two expanding cones of darkness, the larger one astern. Due to a combination of aberration and Doppler effects the visible stars which remained would be concentrated into a narrowing 'barrel' around the ship with blue stars on the leading edge and red stars on the trailing edge. As the speed built up further all the remaining stars would be concentrated into a narrow circular 'rainbow' in the forward hemisphere.

In fact, nothing so clear-cut would be seen. Real stars emit over a wide range of wavelengths, and there are some stars which emit more strongly in the infrared or ultraviolet than in the visible; these are the cool red stars and the hot O- and B-type stars, respectively. A star like the Sun will still be visible when its peak wavelength is red-shifted into the infrared because its ultraviolet radiation will be shifted into the visible; conversely, if we were approaching a Sun-type star at a large fraction of the speed of light, its infrared radiation would be rendered visible. Admittedly, in either situation, the Sun-type star would appear *fainter* but it would not vanish until high percentages of light speed were attained; even then it could be detected with infrared or ultraviolet sensors.

James D. Wertz* has argued that bright blue supergiant stars should be selected as beacon stars for the rearward hemisphere as, when their radiation is red-shifted, because they emit more strongly in the ultraviolet than in the visible they will actually appear *brighter* than normal at high speeds. Likewise, brilliant red supergiants, such as Betelgeuse in Orion, should be used as forward markers, since their greater proportion of infrared radiation will be shifted into the visible region, so rendering them brighter than when viewed from a stationary starship.

The relativistic interstellar navigator is going to be faced with some curious sights and some interesting problems. He may to some extent have to rely upon sensors of invisible radiation to allow him to detect particular stars, but there should be no insuperable problems in his way.

Interstellar Communication

We already possess the capability to communicate over interstellar distances, and on 16 November, 1974, the first major message from Earth was transmitted at a frequency of 2380MHz from the giant 300-metre radio dish of the Arecibo radio observatory, Puerto Rico. It consisted of a basic three-minute message in binary code and was directed towards the globular star cluster M13 (Plate 6) which contains several hundred thousand stars, and which lies at a distance of nearly 7500 parsecs! Despite the phenomenal distance it would, in principle, be possible for a civilization with a similar instrument to receive this signal

*James D. Wertz, *Spaceflight*, vol. 14, no. 6, June 1972.

and attempt to decode it. If they chose to send a reply, of course, it would not arrive until 48,000 years after the sending of our original. Quick-fire repartee would be out of the question.

Communication between ourselves and an established human colony possessing identical equipment would be no problem using existing radio and microwave techniques (although there would be frustrating delays), but communication with a starship would be somewhat more tricky for a number of reasons: the problem of carrying a sufficiently large antenna on the ship; the transmission power requirements; the Doppler effect changing the wavelength of the signal and reducing the rate at which data may be exchanged; the effect of the starship's own exhaust (when under powered flight) on the signal; and so on. None of these problems seems to be insuperable.

The range limit for interstellar communication depends on a number of factors—the diameter of the receiving antenna, the power of the transmitter, and the gain of the transmitting antenna. Gain is a measure of how *directional* the transmitter is; the higher the gain, the

22
An artist's impression of a SEASAT-A spacecraft as it studies the oceans from orbit. The spacecraft will determine if microwave sensors have value in providing information about sea-state and related weather phenomena. The craft orbits the Earth fourteen times a day at about 800km above the Earth's surface, giving coverage of 95% of the oceans every 36 hours. (*NASA.*)

overleaf
A relativistic interstellar ramjet viewed from one of its small shuttle craft moving at the same velocity as the parent craft. The ramjet utilizes a magnetic scoop to suck up ionized hydrogen gas which is 'burned' as fuel in a fusion rocket of advanced design. This particular ramjet has already visited a distant planetary system and is accelerating at a steady rate of 1g (so that the crew feel their normal 'weight'). It has now attained a velocity two-thirds that of light and at this velocity the appearance of the background stars is markedly altered. The stars appear to be considerably thinned out and, as the observer looks further behind, the number of visible stars becomes few indeed. The effect is accentuated by the red-shift which makes stars appear 'redder' than they really are ; cooler stars become invisible because their light is shifted into the infrared part of the spectrum. It is still possible to see a few blue stars, however, because the very hot stars which emit most strongly in the ultraviolet now appear to be emitting strongly at visible wavelengths.

Although the nebula, which also shows colour variations owing to the high velocity, appears to lie in line with the ramjet, this is an effect of aberration ; in fact, the nebula is now far behind the craft.

Other visible details of the ramjet include the control deck, habitation and payload modules, and the advanced fusion motor.

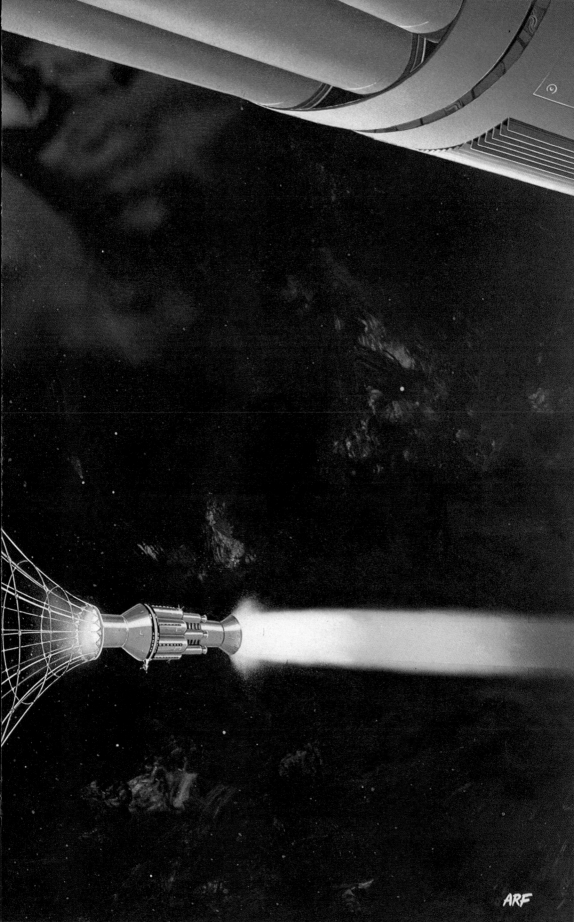

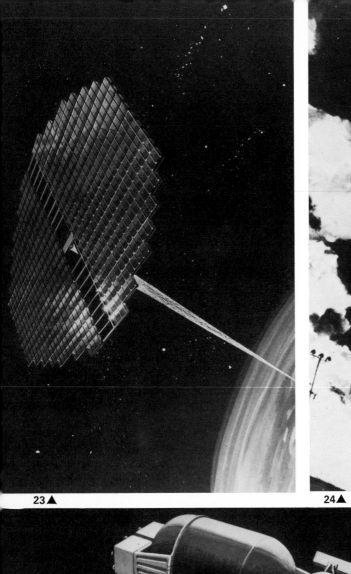

23▲   24▲   25▼

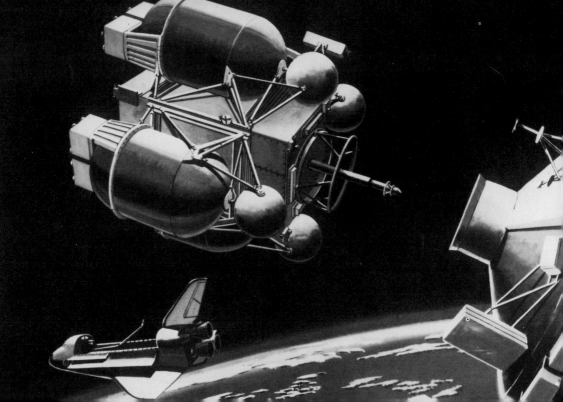

narrower the beam of radiation. For interstellar communication to a specific target there is no point in wasting power by spreading it out further than can be avoided; therefore a high-gain antenna is essential. Factors tending to reduce range include radio noise, generated in space or in the receiver, and bandwidth, the range of frequencies covered by the transmission.

We have to trade the rate of data transmission against range, too. We cannot have it both ways—if we wish to transmit data at a great rate then we must expect the range over which we can do this to be less than the range attained at a more modest rate. The data is transmitted in binary code—a code whereby numerical information can be transmitted in the form of numbers made up out of only two digits (1 and 0)—and decoded at the receiving end. The rate of data transmission is expressed in terms of the number of bits per second, where one bit is one binary digit (0 or 1); we talk therefore of the 'bit rate'.

The progress which has been achieved in interplanetary communication over the past two decades has been quite staggering. Mariner 4 flew past Mars in 1965 and sent back a number of photographs; each picture was divided up into 40,000 tiny elements (or 'pixels')—just as a newspaper photograph is divided into large numbers of tiny dots—each of which was represented by a six-digit binary number corresponding to one of 128 possible shades of grey. Each point required six bits of information so that the total number of bits per picture was

23
Artist's impression of one of several configurations of space-based power conversion systems currently being considered. Placed in geosynchronous orbit, this system could provide 5000 to 10,000 megawatts by collecting energy from the Sun and beaming it down to Earth in the form of microwaves. (*NASA.*)

24
Artist's impression of the blastoff of a Heavy Lift Launch Vehicle currently under consideration by NASA. This vehicle has a greatly increased economy and payload flexibility over the basic Space Shuttle for the heavy payloads required in the building of large structures in space; it employs all the propulsion systems and the external tank of the Space Shuttle but does not incorporate the orbiter itself. (*NASA.*)

25
A teleoperator retrieval system approaching a satellite in order to dock with it and adjust its orbit. This system is expected to be used in conjunction with the Space Shuttle in the early 1980s for the retrieval, surveillance, manoeuvring and stabilization of payloads in low Earth orbit. (*NASA.*)

240,000. At a rate of 8.3 bits per second it took *eight hours* to transmit one picture. The Voyager spacecraft, launched in 1977 and scheduled to reach Jupiter in 1979, are expected to have a bit rate, from Jupiter, of 115,200 bits per second. This bit rate is about fourteen thousand times better than that attained with Mariner 4, and over a much greater range, too. A picture from Jupiter will contain over 5 million bits of information (for comparison, the entire contents of this book—suitably coded—could be expressed in about half that number of bits) and will be transmitted with no more than 28 watts of power—not enough to light satisfactorily an average table-lamp. Such is the measure of progress in communications technology.

The only problem likely to cause much difficulty—apart from the inevitable time-delay in messages—is the effect on the signal of the clouds of ionized gas emitted from the rocket exhaust during powered flight (all the time, for the $1g$ spacecraft). Those of you who remember the re-entry of the Apollo spacecraft on their return to Earth will recall the tense period of a few minutes when communication was lost through the build-up of hot ionized gas around the spacecraft as its external temperature soared due to atmospheric friction. An interstellar rocket almost certainly will create its own clouds of ionized gas which—during the acceleration phase—will be interposed between the spacecraft and the Earth, interfering with microwave signals.

Ways around this can be found. One possibility might be to shut down the motors periodically to allow communications. Alternatively, another kind of system may be employed during the boost phase, an optical or infrared laser, for example, which—because it operates at a much shorter wavelength than microwaves—can be made much more directional. R. L. Forward has suggested that a 1-kilowatt laser beam, transmitted *via* a 100-metre-diameter reflector, would be detectable at the distance of Alpha Centauri by means of a quite small telescope.

Although some work will have to be done on efficient systems it does seem as if communication with starships will not pose overwhelming problems. In this chapter we have looked at some of the general questions associated with interstellar flight; in the next one we shall examine the crucial problem of the means of propulsion which may be available to power the starships of the future.

7 The Power to Reach the Stars

Are there any foreseeable propulsion systems which would be capable of driving starships up to appreciable fractions of the velocity of light, so allowing interstellar missions to be undertaken in reasonable periods of time—decades rather than centuries or millennia? Without stretching the bounds of present scientific knowledge and without stretching technological developments to ridiculous extremes it seems quite possible that such means may well become available to us before too long. In this chapter we shall examine some of the possibilities which are being discussed. We cannot say that all, or even any of them, will prove to be practicable, but the fact that there are so many possibilities already being debated suggests that the problem is unlikely to remain unsolved for long.

None of the systems described in previous chapters is suitable for interstellar flight—the solar sail would literally run out of wind at large distances from the Sun, as indeed would the solar-electric ion rocket (SEPS); the nuclear rocket which uses the reactor to heat up a working fluid such as liquid hydrogen is equally incapable of attaining anything remotely approaching the necessary cruise velocity. What is required is something new.

The Nuclear-Electric Ion Rocket

An ion rocket deriving its electrical power from a nuclear generator (a fission reactor) would be freed of the range limitation of the SEPS system but, even if an improvement in exhaust velocity of a factor of ten or twenty could be attained over present ion rockets, this would yield exhaust velocities only of the order of 1000km/s. To attain a cruise velocity of 10% of the speed of light with such a system would require a mass ratio of ten million million. The more modest target of 1% of the speed of light could be attained with a mass ratio of only 20, but this velocity would imply long travel times—a thousand years for a nominal three-parsec mission.

To achieve the maximum mass ratio, a nuclear-electric rocket would have to consume much of its structure by converting it into ions. One might envisage some future maniacal 'Captain Ahab' ionizing members of his crew in order to attain that last minute additional quantity of acceleration! On the face of it the system does not seem very promising so far as 'fast' starships are concerned.

The Fusion Rocket

Much effort is being expended here on Earth in tackling the problem of the continuous production of power from fusion reactions. So far our attempts to try to harness the power source which drives the Sun have met with success only in the hydrogen bomb—a somewhat uncontrollable source of energy. If it were possible to produce energy by hydrogen fusion reactions under controlled conditions then, since water contains hydrogen, and seawater is abundant, the current 'energy crisis' would vanish at a stroke. Furthermore fusion power, unlike fission power (which in any case relies on scarce fissile materials), produces no long-lived radioactive by-products.

The problems are enormous. For the reactions currently being investigated to proceed, the reacting plasma (ionized gas) must be at a temperature of over 100 million Kelvins and must be contained in a confined space. The only method available is to use a 'magnetic bottle' whereby the plasma is trapped within a powerful magnetic field, and it is here that the main problem arises. The bottle must be leak-free and stable; any instability quickly leads to the destruction of the bottle and the loss of the hot plasma.

Despite thirty years' research it has not yet proved possible to contain the hot plasma for more than a minute fraction of a second and there is clearly a long way to go before fusion power becomes a reality. Nevertheless, few would doubt that it *is* attainable given sufficient funding, research and development.

Would fusion power provide the key to interstellar propulsion? One advantage that the fusion rocket would have over the fusion reactor is that the plasma does not have to be contained in the same way. After all, a rocket motor is fundamentally a *leaky* bottle, the exhaust gases being allowed to escape in a specified direction; a fusion rocket would require a magnetic bottle which leaked at one end!

The characteristic exhaust velocity which is commonly quoted for a fusion rocket is 10,000km/s, about 3% of the speed of light. A mass ratio of 20 would be sufficient to attain a cruise velocity of one tenth of the speed of light and this would bring Proxima Centauri within about 43 years' flight time, although a three-parsec mission would require 100 years in space, which is too long for a manned flight. Of course a mass ratio of 20 squared—i.e., 400—would be required for a mission which slowed down and landed at the target, and for a return flight—if the vehicle had to carry all the fuel for the outward and return leg from the Earth—the total mass ratio would be 160,000! Return flight will have to rely upon refuelling at the target. Building the vehicle in stages will, of course, greatly improve the position, but there are limits to this process since each stage would require its own fusion motor.

There are other problems too. To be effective, a rocket must expel mass, but in many fusion reactions less than 20% of the energy released comes out in the form of fast-moving massive particles, the remainder escaping in the form of short-wave radiation (X-rays and the like). Apart from being wasteful, this poses a hazard to life and suggests that massive shielding would be required on manned starships which, in turn, would make it hard to achieve good mass ratios. A reaction must also be selected which minimizes the proportion of neutrons released as these too would imply a need for massive shielding.

Much remains to be answered concerning the feasibility of the fusion rocket, but there seems every likelihood that it may one day make a contribution to Man's exploration of the stars.

The Nuclear Pulse Rocket

In 1891 Hermann Ganswindt, a prolific German inventor, suggested it might be possible to build a rocket propelled by a series of explosions, each of which supplied an impulse to the rocket. It may sound bizarre and hazardous, but the idea is not as crazy as it might seem; after all, most forms of transport on Earth are propelled by the internal combustion engine, which derives its power from the regular series of explosions each of which pushes down a piston which in turn rotates the crankshaft. Such a motor is capable of producing smooth acceleration because the individual explosions—which are relatively small—take place in very rapid sequence. A suitably damped and carefully regulated series of explosions should be capable of accelerating a rocket vehicle in a smooth fashion.

The concept was examined seriously in the nineteen-fifties under the heading of Project Orion. Originally conceived in 1955 by Dr Stanislaw Ulam at the Los Alamos Scientific Laboratory, New Mexico, the scheme was to lob a series of atomic bombs out of the back of the vehicle, part of the momentum of the explosion debris being absorbed by a pusher plate attached to the spacecraft by some kind of shock absorbers which imparted a relatively smooth acceleration to the craft. There were many potential problems. For example, the pusher plate would have to be protected by some kind of ablative material which, as it vaporized, carried away the tremendous heat of the explosion (otherwise the pusher plate would melt); and the shock-absorbing mechanism would have to be elaborate and, above all, massive (the pusher plate itself would be of the order of thousands of tonnes).

Since the scheme envisaged using fission bombs, the explosions could not be small; for a certain critical mass of fissile material is required. Fission reactions are 'dirty', releasing harmful radioactive products, and such a vehicle could not be allowed to fly in the Earth's atmosphere. Even in space, massive shielding would be required for the crew and onboard instrumentation. Although it may seem about as subtle as kicking a can along the road, and although the technique would be very wasteful of energy (only a small proportion of the energy released in the explosion would go into accelerating the spacecraft) the system offers in principle a far more powerful means of propulsion than the type of rocket discussed in Chapter 5.

The US Department of Defense became very interested in the project and supplied funding for preliminary research and development. Although much of the project is still classified it is believed that small test models were flown using conventional explosives in about 1969. NASA became interested in the project in the early nineteen-sixties as a possible means of propelling manned interplanetary flights, but it is believed that the project was abandoned short of nuclear testing in the face of the nuclear test-ban treaty.

The thermonuclear fusion reaction releases more energy and is cleaner in the sense of not producing long-lived radioactive products. However, the conventional 'hydrogen bomb' is detonated by means of a fission device and so still produces radioactive fall-out. Freeman J. Dyson has done some calculations on a hypothetical fusion pulse rocket. The system discussed had an initial mass of 400,000 tonnes (much the same as a modern supertanker), two thirds of which was made up of 'propellant' in the form of 300,000 hydrogen bombs, each of a little under one megaton yield. After allowing for the structure, 45,000 tonnes would be available for payload—sufficient to make up a small space colony. By having one detonation every three seconds, $1g$ acceleration could be maintained for ten days leading to a final velocity of about 3% of the speed of light. The flight time of 300 years for a nominal three parsec mission would have to be doubled if half the bombs were retained for deceleration at the target; such a mission is out of the question in an individual lifetime, but would be feasible for the space ark concept.

In such crude terms the idea seems wholly impractical, but in recent years the concept of the nuclear pulse rocket has been greatly refined and it probably represents the only practicable mode of interstellar propulsion which could be realized next century by the development of existing and foreseeable technology. The first full-scale starship study, utilizing this technique, was carried out over the period 1973–1978 by a study group of the British Interplanetary Society and the proposed mission, known as Project Daedalus, is discussed later in this chapter.

The key to success lies in producing very large numbers of (by today's standards) very small thermonuclear explosions by using powerful lasers or focussed beams of relativistic electrons to detonate small pellets of fuel. In either case the idea is to focus large amounts of energy onto the very small surface area of a pellet, the resulting pressure causing the pellet to implode and initiate the desired fusion reaction. F. Winterberg, who has published some of the fundamental papers on this technique, favours the use of intense pulsed electron beams as being more efficient in supplying energy to the pellets—and this type of approach has been adopted in the Daedalus project—but recent research has shown the feasibility of developing pulsed lasers (i.e., lasers having outputs consisting of a series of pulses of extremely short duration) capable of delivering the requisite quantities of energy and these, in the end, may prove to be the most suitable triggers.

As with the fusion rocket, it is necessary to select a reaction which does not release a flood of neutrons. The reaction favoured by the Daedalus group involves deuterium (D)—a 'heavy' version of hydrogen, the nucleus of which consists of one positively charged proton and one neutron—and helium 3 (^{3}He), the rare 'lightweight' isotope of helium, having two protons and one neutron (normal helium has two of each and is denoted by ^{4}He). The fusion of D + ^{3}He produces ^{4}He + a proton + energy. The fuel pellets consist of small spheres, a few centimetres in diameter, of deuterium and helium 3. These are injected towards the centre of a magnetic field inside a reaction chamber and, when they reach the appropriate point, are bombarded by the electron beams which render fusion possible. The expanding ball of plasma resulting from the explosion interacts with the magnetic field in such a way that the reaction products are expelled from the reaction chamber and the chamber itself is pushed forward. The designed explosion rate is 250 per second, leading to a much smoother acceleration than would the old Orion system.

One major disadvantage of this choice of fuel is that helium 3 is extremely rare in nature and, since tens of thousands of tonnes of it would be required for each spacecraft, the Daedalus study group have suggested that it might be 'mined' from the atmosphere of Jupiter.

Winterberg has suggested an alternative reaction which does not rely on the rare ^{3}He isotope, and utilizes instead the fusion of hydrogen and boron to form helium. Unfortunately this requires much higher ignition temperatures than the Daedalus reaction, but he has suggested that the way round this problem may be to use 'staged nuclear microexplosions' whereby a small explosion which is itself easily triggered provides the energy to set off another of greater yield. In a series of steps, the difficult hydrogen–boron reaction would eventually be triggered.

Nuclear pulse rockets offer the prospect of attaining velocities in excess of one tenth of the speed of light (the Daedalus probe has a designed cruise velocity of 12% to 13% of light speed) without the need for crippling mass ratios; the Daedalus craft, which would not be decelerated at the target system, would be a two-stage vehicle with an overall mass ratio of only 12 or 13. Considerable attention has already been devoted to the technique, and, although when they do fly the precise details will doubtless be different from what has been described here, the fact remains that nuclear pulse rockets appear to require only enhanced development of twentieth-century technology to become a practical reality. The system must be a strong contender for the first true interstellar mission.

The Interstellar Ramjet

As we have already seen, the problem with rockets is that they have to carry their fuel with them, and waste much of the energy produced in accelerating the remaining fuel. The ideal solution would be a rocket which collected its fuel as it went along, and it is this promise which is held out by the interstellar ramjet (ISR), a mode of propulsion which—if it can ever be made to work—will open the door to true interstellar flight on a very large scale.

As we saw earlier, the space between the stars is filled with a tenuous mixture of gas (and dust), mostly in the form of hydrogen and helium at extraordinarily low density (about 10^{-21}kg/m³ on average; i.e. less than one atom per cubic centimetre); in denser clouds, however, the density may be a thousand times higher than this. The gas may be used as fuel for some kind of fusion motor provided that the ramjet can scoop up sufficient quantities to sustain a suitable reaction. The original discussion of this idea was published in 1960 by R. W. Bussard,* who suggested that a 1000-tonne spacecraft would be able to sustain 1g acceleration through interstellar space provided it had an intake area of 10,000 square kilométres in a high density region, and 10,000,000 square kilometres in a low density region (circular intakes of these areas would have diameters of about 100 kilometres and 3000 kilometres, respectively). Of course, the physical structure of the 'scoop' would not be so large; rather the scoop would consist of a structure which generated a powerful magnetic field which influenced ionized gas within the required radius of the ramjet and funnelled it into the reactor system. The ramjet may be visualized as a cosmic vacuum cleaner, sucking up its fuel as it proceeds on its way.

It is generally felt that the original estimates were rather optimistic, and some recent calculations have tended to be pessimistic about the prospect of the ISR attaining anything like the ideal 1g acceleration. Nevertheless, the ramjet concept is such an elegant one that it would be tragic if it could not be developed to allow the possibility of completing major interstellar flights within a human lifespan. It is a concept which deserves to succeed.

The ramjet has to be moving at a fair speed before it becomes effective at scooping up sufficient material to sustain fusion, and will have to be given an initial boost by a more 'conventional' means of propulsion such as the nuclear pulse rocket or the fusion rocket. It will be essential to carry sufficient fuel at least for the initial acceleration phase. However, the ramjet becomes effective at small fractions of the speed of light; according to Alan Bond† it reaches 50% of full thrust at 2% of

*R. W. Bussard, 'Galactic Matter and Interstellar Flight', in *Astronautica Acta*, vol 6, pp 179–194, 1960.

†Alan Bond, 'Problems of Interstellar Propulsion', *Spaceflight*, vol 13, no 7, July 1971.

light speed. The initial boost is well within the capability of either of the other two propulsion systems. A fusion propulsion system would require an initial mass ratio of about 2 to attain something like 2% of light speed; thereafter the ramjet would come into its own, accelerating steadily to higher and higher velocities.

As speeds approached that of light, relativistic effects would come into play. Because of length contraction, distances in space would appear shorter to the crew and *therefore* the number of hydrogen atoms per square metre would appear to increase. So far as the spacecraft were concerned the rate of interstellar fuel flow would increase, allowing the efficiency of the engine to be reduced for the same acceleration, or allowing a higher acceleration rate to be achieved. By the same token the same rate of acceleration could be maintained in a more rarefied region of space. So far as the ramjet is concerned, the faster the better.

The ramscoop itself could provide a means of decelerating the craft. With the fusion motor switched off, the magnetic field would be employed to 'reflect' the interstellar gas, so slowing down the spacecraft by drag. In effect, the ramscoop would become a 'parachute'. Of course, the effectiveness of the parachute would diminish as the speed of the ramjet decreased, and it would still be essential to use some kind of conventional rocket to make the final velocity reduction from a few per cent of light velocity to zero. A ramscoop, or an electrostatically charged 'sail', could be used as a means of braking any kind of interstellar craft.

It must be admitted that the theoretical and technical problems associated with designing a fusion system to operate on interstellar gas are formidable. The proton–proton reaction by means of which the Sun produces energy from raw hydrogen would be ideal, but operates effectively only at very high densities such as are found in the core of the Sun. If it is not possible to harness a reaction of this kind then it will be essential to use a fusion reaction such as those being considered for the nuclear pulse rocket, which rely upon commodities which are much rarer in the interstellar gas than is hydrogen, and the difficulty will arise in obtaining sufficient *suitable* fuel.

More readily attainable may be the ram-augmented interstellar rocket (RAIR), in which the interstellar gas is not used directly as fuel, but simply as reaction mass. In the RAIR the ionized interstellar gas would be accelerated by an electric field generated by means of a nuclear reactor on board the starship, and expelled from the rear of the vehicle to provide forward thrust. Although the RAIR—which is in effect a self-fuelling ion rocket—cannot be as effective as the true ramjet, it still has considerable potential.

The basic concept of the interstellar ramjet, although it may not be attained in practice for some considerable time, nevertheless offers the best current hope for long-range interstellar travel. Its essential elements are, a ramscoop exerting its magnetic influence for tens of thousands of kilometres ahead of the spacecraft and around its line of travel, a fusion motor which utilizes this fuel, and a boost rocket which accelerates the spacecraft to a sufficiently high velocity for the ramjet effect to take over. The major difficulties lie in the means of producing the field, of building-in sufficient structural strength to the scoop components, and in the design and operation of a fusion reactor capable of sustaining a workable reaction. These are massive problems which I have every confidence will be overcome.

The Laser-Photon Sail

We have already mentioned the idea of the solar sail, a device propelled by radiation pressure from the Sun, and which is under serious consideration as a means of interplanetary propulsion. Since the intensity of sunlight diminishes rapidly with increasing distance, such a sail could not be used for interstellar journeys.

Robert L. Forward has suggested that a powerful bank of lasers— based in orbit around the Sun and drawing their energy from it— could be used to project a beam of light onto a sail and so accelerate the spacecraft to which the sail is attached. The laser arrays would have to be very large (Forward suggests 250 kilometres across) in order that the beam would not diverge too far at distances of several parsecs. Close to the Sun, in the initial stages of the journey, a smaller bank of lasers could be used, and the mission could commence before the complete bank was ready, additional units being added as the starship got further and further away. As the spacecraft sailed down the beam it would be provided with an ideal navigation device in the form of the beam itself; since this would be directed towards the target, all the spacecraft need do is to keep within it.

One drawback of this mode of propulsion should be obvious—how can the starship be halted at the end of its journey? This does not matter for a fly-by mission, but if a landing or extended exploration is planned some means of deceleration will be required. A ramscoop might be a possibility, but some kind of conventional rocket motor would still be required for the final stages. One ingenious solution* proposes that, when the craft has been accelerated to the appropriate cruise velocity, the lasers be switched off and the craft would then extend long charged wires from its structure which would interact with the galactic magnetic field to swing the craft round through a wide semicircle beyond the target star. When the spacecraft was still beyond the target but heading back towards the Earth, the laser bank would again be switched on to decelerate the probe prior to reaching its target.

I doubt that we shall see interstellar spacecraft propelled in quite

*See R. L. Forward, 'A Programme for Interstellar Exploration', *Journal of the British Interplanetary Society* vol 29 no 10, 1976.

this fashion, but it is worth noting that great strides are being made in the development of lasers as a means of transmitting power in space. The next decades should see practical systems of this kind, and it is not beyond the bounds of possibility that power may be transmitted from the Solar System to interstellar spacecraft. For example, interstellar arks might be kept supplied in this way while on their long and tedious journeys, although, if some future 'government' decided to withdraw funding (and therefore power transmission) from the project in an economy drive, then the outlook would be bleak indeed for the inhabitants of the ark.

Daniel P. Whitmire and A. A. Jackson IV have recently discussed a thoroughly ingenious hybrid proposal,* the laser-powered interstellar ramjet (LPR) which utilizes a ramscoop to collect hydrogen to be used as reaction mass, and uses the energy collected from a powerful focussed laser beam to energize this mass. Ignoring the problem of focussing all the energy in the beam onto a starship at a range of several light years (and these problems are formidable), calculations suggest that in principle the system should be more efficient than the laser sail (LPV) which utilizes the momentum of photons. Because the vehicle will be using the energy of the laser beam, high-energy photons will be required, suggesting the need for using X-ray lasers. At relatively low speeds (up to 14% of light speed), it should be more efficient than the 'conventional' interstellar ramjet (ISR), but above that speed the ISR becomes more effective.

The beauty of this scheme is that the energy of the laser beam can be used also to decelerate the receding spacecraft. The spacecraft still scoops up hydrogen as it goes along (and this in itself provides a certain amount of drag) but the thrust of the engines is reversed. Only in the final stages of deceleration would a conventional fuel supply be required.

Even more intriguing is the possibility of using the same laser beam for the return journey. In fact the LPR should be *more efficient* when flying into the beam than receding from it, as in this mode it absorbs more photon energy per second (the laser energy is 'blue-shifted' to higher frequencies and therefore to higher energies). Whitmire and Jackson estimate that travelling into the beam the LPR can actually be more effective than the ISR at all speeds! Two-way journeys can be accomplished with the single beam, and the onboard fuel requirements are slight; with such a system the crippling mass ratios associated with more conventional interstellar rockets could be disposed of. Furthermore, the LPR does not require an advanced form of fusion engine to 'burn' the hydrogen fuel.

The ideal set-up would be to have a network of laser beams set up at 'Federation' starbases, so that interstellar ships would always travel into

*Daniel P. Whitmire and A. A. Jackson IV, *Journal of the British Interplanetary Society*, vol 30, pp 223–226, 1977.

the beams (no navigation problems either). In time one could envisage a whole system of interstellar links with an 'up' beam and a 'down' beam, rather like the intercity railways of today.

More Esoteric Possibilities (1): Antimatter

'Antimatter' is a word which catches the imagination and which appears to have an almost mystical connotation. In fact, it is not nearly so mysterious as it sounds. In certain nuclear reactions, antiparticles are produced which have opposite properties to ordinary particles. Sub-atomic particles can be described by certain basic properties such as mass, charge and spin (we can visualize these particles as spinning clockwise or anticlockwise). For example, a normal particle such as a proton has one unit of mass, one unit of positive charge, and a particular value of spin; its antiparticle would have the same mass, one unit of *negative* charge and the opposite value of spin. Again, the antiparticle of the electron (negative charge) is the positron (of positive charge). This is a rather simple view of subatomic particles, but it will suffice.

Ordinary matter is made up of normal particles such as protons and electrons; antimatter would be made up of antiparticles such as anti-protons and positrons. If antimatter were to exist in bulk then it would look just like ordinary matter; but in our part of the Universe, at least, antimatter exists only on the scale of subatomic physics. In the laboratory we have been able to generate antiparticles, but no quantities of bulk material.

The crucial point is that when a particle meets its antiparticle they annihilate each other totally, releasing their masses as energy in the form of gamma radiation. If a solid lump of antimatter were to meet an equal lump of ordinary matter then, again, the outcome would be total annihilation, the 100% efficient conversion of matter into energy. Does this offer up the prospect of the ultimate in rocket propulsion? The picture is not as rosy as it appears at first glance, for about 50% of the energy is released in the form of neutrinos, which will penetrate any-thing and everything and which cannot be directed by electric or magnetic fields, while most of the rest is in the form of gamma rays. Attempts to dream up suitable gamma ray reflectors to turn some of this energy into useful thrust have been singularly lacking in success.

Perhaps some solution will be found, but it may be more effective to use only a small proportion of antimatter (instead of a 50:50 mix of matter and antimatter) which, by annihilating an equal amount of ordinary matter, will release energy which will expel the remainder of the propellant and so produce useful thrust. If only 5% of the total quantity of propellant is converted into energy, this would still be a more efficient process than what goes on inside the Sun (where only 0.7% of the reacting matter is converted into energy), and if only 10% of *that* energy were transferred to the propellant we should still have a very effective rocket.

R. L. Forward has discussed a conceptual tiny interstellar probe of 10-kilogram payload launched by a two-stage vehicle utilizing matter and antimatter in the ratio 40:1. With a launch mass of 256 kilograms (the second stage plus payload making up 51 kilograms of the total), a final velocity of one third of the speed of light could be achieved. Of course this calculation ignores the mass of the vehicle structure, which would probably be considerable, and a vastly more massive starship would be required to achieve the economy of scale necessary in a practical situation. Nevertheless, the figures are impressive, although for long journeys such a system would not be as effective as the ideal interstellar ramjet.

The production of sufficient quantities of antimatter is a problem which lies well beyond existing technology, and the problems of handling and storage are equally intractable (although they would appear to have been solved by the designers of the Starship *Enterprise*). After all, you cannot manufacture a lump of antimatter and then put it in a box made of conventional matter; it would destroy the box and everything else around in a devastating explosion. The only means of handling bulk quantities of antimatter would be by the precise manipulation of powerful fields, and the consequences of a power cut are unthinkable.

The potential is enormous but the technical problems are over-whelming. They may be solved one day, but I feel this is one mode of propulsion which lies a long way ahead in time.*

More Esoteric Possibilities (2): The Photon Rocket

The idea of the photon rocket is an interesting one. A vehicle propelled by raw electromagnetic radiation has the highest possible exhaust velocity, the velocity of light itself. Since exhaust velocity is one of the factors which determine the final velocity of a rocket, such a prospect is surely appealing.

We find, however, that we are faced with problems similar to those which we have just discussed in examining the matter–antimatter rocket. For each Newton of thrust—assuming it operates at 100% efficiency—the rocket will consume power at the rate of 300 megawatts, and to attain the 'ideal' 1g acceleration will require 3000 megawatts per kilogram of spacecraft. The only way to get final velocities which are a useful fraction of the speed of light without incurring ludicrous mass ratios is by the conversion of mass to energy at an efficiency approaching 100%, and the only process which can achieve this is the matter–antimatter interaction with all its concommitant problems. It may be that future physicists will find other means of producing energy at high efficiency and, if so, the photon rocket may become feasible.

*In making this remark I am probably making my only constructive contribution to the development of interstellar propulsion. For the one thing which is almost certain is that dogmatic statements like this are usually very wide of the mark—in which case the realization of the matter-antimatter drive should be greatly facilitated by my comment.

Given a suitable energy source there remain the twin problems of how to radiate all or most of the power in the right direction, and how to protect the crew from the effects of radiation and waste heat. The design of a suitable reflector for high energy radiation remains a problem. The photon rocket offers the potential of very high final velocities, but its development lies well beyond the bounds of existing technology.*

Project Daedalus

These are some of the propulsion systems which may be instrumental in taking Man to the stars. Perhaps new technological and scientific developments will lead to alternative and more effective means but, as things stand at the moment, the ideal system for very long journeys would seem to be the interstellar ramjet, while at closer range—given sufficiently powerful and accurate laser systems—the laser-powered ramjet looks most promising. Both of these techniques lie well beyond us at the moment, but the one hopeful approach which does appear to lie not too far beyond our grasp is undoubtedly the nuclear pulse rocket. This may well prove to be the system which drives the first interstellar probe or, indeed, the first insterstellar ark, should the space ark concept turn out to be acceptable to humanity.

Since February 1973, the Project Daedalus Study Group of the British Interplanetary Society has devoted rather more than 10,000 man-hours to producing a detailed specification of a possible interstellar mission utilizing the nuclear pulse rocket. In what is probably the most detailed investigation of its kind, the group has examined all aspects of the mission, from the selection of the target, the specification of the vehicle and its payload, to the question of navigation and communication with the probe throughout the mission. Only a brief outline of the proposal can be given here, but full details are contained in the report of the Study Group,† published in the spring of 1978.

The target star selected is Barnard's Star, a dull red star of spectral type M5, less than one two-thousandth of the Sun's luminosity; so that, despite the fact that it lies less than 2 parsecs away, it is too faint to be seen without telescopic aid. However, as we have already said, it is the one star for which there is good evidence of the existence of a planetary system. The mission would take the form of an undecelerated fly-through, but as the main spacecraft hurtled towards the target system a succession of smaller probes would be released to investigate the individual planets.

---

*A former colleague, D. A. Robinson, once remarked (tongue in cheek) that perhaps quasars, which are point-like sources of radiation exhibiting strong red-shifts, were the exhausts of photon rockets—in which case it would seem as if some more advanced beings from our Galaxy have already developed this technique and are shooting away in all directions to colonize the rest of the Universe.

†*Project Daedalus*, British Interplanetary Society, April, 1978.

THE POWER TO REACH THE STARS

The propulsion system would be a nuclear pulse rocket using the deuterium–helium 3 reaction which we have already described. A two-stage vehicle is envisaged, having an initial mass of 54,000 tonnes, of which 50,000 tonnes would be propellants, the total payload—including a set of individual sub-probes to be dispersed as the target system was approached—making up about 500 tonnes. The designed flight time to Barnard's Star is about 50 years, and this requires a cruise velocity of about one-eighth of the speed of light; this would be achieved after an acceleration period lasting nearly four years.

Considerable thought has been given to the question of erosion protection, and it is felt that beryllium shielding 7mm thick should suffice for the interstellar coast phase. After entering the system of Barnard's Star, dust and debris would constitute a much greater hazard to a vehicle travelling at such high speed and it is proposed to deal with this by deploying a cloud of particles some 200km ahead of the vehicle to dispose of large particles. The small dust which survives impact with the cloud should be handled adequately by the beryllium shield.

30,000 tonnes of helium 3 would have to be gathered from the helium-rich atmosphere of Jupiter by means of separation plants floated in the atmosphere beneath giant hot-air balloons. This in itself would be a massive technological project.

The navigation system will rely largely on the measurement of stellar positions and the highest accuracies will be demanded at the time of the mid-course correction manoeuvre and on the approach to the target system when the sub-probes will have to be targeted and launched. The probe will have to carry instrumentation capable of detecting planets while the probe is still several years out from the system, so that the launching of probes can be carried out with the minimum expenditure of fuel. It is envisaged that the payload will include telescopes of the order of 5 metres aperture (i.e., as big as the largest present Earth-based optical instruments). The possible scientific payload has been examined in some detail, the main areas of likely interest being the interstellar medium during the flight, the physics of the target star itself and its environment, and planetology (the study of the planets in the Barnard's Star system). Data will be relayed during the powered flight phase by means of an infrared laser system, while in the coast phase and at the target system a more conventional microwave link will be employed.

Allowing for twenty years for the design and construction of the vehicle, fifty years in transit and six to nine years' transmission of information back to the Solar System, the Daedalus project would occupy nearly eighty years in all.

No one can say that this *is* the way in which the first true interstellar mission will be undertaken, but there is nothing in Project Daedalus which is inherently absurd; the whole project has a quite convincing ring to it. To those who suggest that interstellar flight is impossible, the answer is: 'Nonsense!' The Pioneer probes have already achieved Solar System escape velocity, there is considerable serious published work on potential propulsion systems capable of speeds of a good fraction of the velocity of light, and there must be every chance that by the middle of the twenty-first century a practicable system will be developed.

We may not be as close in time to the first starship as was Tsiolkovskii to the first Sputnik—but I doubt very much that we are as far removed from it as was Jules Verne from Apollo 8.

26
A solar power station in the process of construction nearly 60,000km above the Earth. (*NASA.*)

27
The 30m panels stretching out like wings on each side of this ion rocket contain thousands of solar cells which convert sunlight into electricity which, in turn, powers the ion engines. It is proposed that this solar electric propulsion stage be used for a number of Earth-orbit and interplanetary missions. (*NASA.*)

28
Final stages in a mission to Jupiter whose proposed launch date is in January 1982. Here the probe portion of the spacecraft is launched towards the planet, an event scheduled for late summer 1984. The probe, launched from the orbiter 56 days before the latter's arrival in Jovian orbit, will enter the atmosphere on the sunlit side, taking atmospheric measurements during its 30-minute descent. The orbiter itself would circle Jupiter for at least 20 months, studying the planet, its larger satellites and the entire Jovian environment. (*NASA.*)

29
The descent through Jupiter's atmosphere of the probe shown in the previous plate. The red-hot nose cone has just separated from the probe portion. The probe itself, braked by a conventional parachute, is taking the first direct samples of the Jovian atmosphere. (*NASA.*)

30
Artist's impression of a Mars Rover exploring, in 1984, the 4000km-long, 6·5km-deep canyon on Mars, Valles Marineris. The Rovers, currently under study, have been described as intelligent in that they would possess a rudimentary ability to learn: equipped with proximity sensors, stereo cameras, laser-ranging instruments and advanced computers, they would traverse at least 100km of the Martian surface independent of instructions from Earth. Each of the proposed two Rovers, about the size of a large desk, would carry over 100kg of scientific instruments to study the Martian surface. Power comes from a 250-watt radioisotope thermoelectric generator at the rear of the Rover. (*NASA.*)

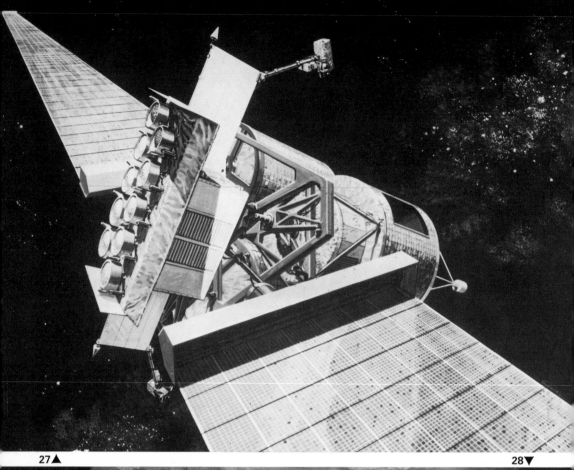

8 Breaking the Rules

The speed of light is an absolute barrier to all space transportation systems. Can we hope to break through this barrier?

It is a central element of Relativity theory that no signal can travel faster than light, and it follows from this that no material object can hope to attain this velocity. The predictions of this theory have been rigorously examined and, to date, the theory has stood up to all the experimental tests to which it has been subjected. For example, it is well established that the mass of a moving object increases with increasing velocity and that this effect increases rapidly as velocities approach close to that of light. At everyday speeds, the effect is too small to be noticed; even the fastest men in history—the Apollo astronauts—had their masses increased by only a few parts in a hundred thousand because of relativistic effects. However, at 87% of the speed of light a body moving relative to an observer on the Earth will have double its rest-mass (the mass which it would have if it were stationary relative to our observer); at 99% of light velocity the mass increase is by a factor of 7; and at about 99.99% of the speed of light the mass of a moving body is increased nearly a hundred times. As the speed approaches closer and closer to that of light, the mass increases towards an infinitely high value, and no matter how much energy we supply to the spacecraft, we cannot propel it exactly *at* the speed of light. Even the conversion to energy of our entire Galaxy would not suffice to propel the dot on this 'i' to the speed of light. In effect, an infinite amount of energy would be required to achieve this.

31
An ion drive spacecraft chasing and studying Halley's Comet during the comet's next approach to Earth, in 1986. The ion drive (solar electric propulsion system, SEPS) uses a mercury-fuelled ion engine boosted by solar panels of dimensions up to 8m × 40m. (*NASA.*)

If travel *at* the speed of light is impossible, it would seem to be self-evident that we cannot hope to travel *faster* than light. It therefore looks as though our descendants, if they wish to travel over interstellar distances, may be able to do so at near-light speeds, utilizing the time dilation effect, but having to accept the penalty of never being able to return to their own time on Earth.

Is the first sentence of the preceding paragraph as obvious as it seems? Common sense suggests that it must be so. After all, if I am driving along a road at 40kph, and I wish instead to drive at 50kph, then in accelerating to the higher speed I must at some stage be moving at 45kph. The world of atomic and nuclear physics does not necessarily adhere to such 'self-evident' rules. Consider the simple model of the hydrogen atom, in which we visualize a nucleus (made up of one proton) around which there revolves one electron. The electron is allowed to move only in certain specified orbits, each of which corresponds to a different energy level, but it may move from a high energy state to a low energy state without occupying any of the intermediate levels. Or, again, there is the phenomenon of 'quantum-mechanical tunnelling' whereby electrons which have insufficient energy to get over an electrical barrier may nevertheless get from one side of it to the other. It is as if in some way they have managed to 'tunnel through' the barrier. The phenomenon is well established and is used in commercial electronic devices.

Could we devise a means of crossing the 'light barrier' without requiring infinite amounts of energy? Curiously enough, in a sense the Special Theory of Relativity does not really preclude the possibility of travelling faster than light; rather it implies that no material object can travel *at* the speed of light. The equations describing the increase of mass with speed allow of the existence of particles which must always travel faster than light and which cannot *slow down to* light speed. The possibility of faster-than-light particles, known as 'tachyons', has been discussed by a number of physicists. These hypothetical entities would possess 'imaginary mass' but real energy and momentum; their energy and mass would increase with decreasing velocity and would become infinitely large at the speed of light. We can imagine two types of matter on opposite sides of the 'light barrier'—ordinary matter which cannot be accelerated up to the speed of light, and tachyon matter which cannot be decelerated to light speed.

As yet there is no proof of the physical existence of such particles. The fact that an entity is a mathematical possibility does not necessarily imply that it must exist (likewise the fact that something has 'imaginary' properties does not mean that it cannot exist). Attempts are being made to detect tachyons, as yet without success, although some cosmic-ray experiments carried out in 1973 by Clay and Couch of the University of Adelaide did produce curious results which could conceivably be explained by tachyons.

If tachyons exist it may be possible one day to use them to obtain or to communicate information faster than light, but it is difficult to see how faster-than-light (or 'super-relativistic') travel for *Homo sapiens* can be achieved.

Communication with the Earth would be a problem for the super-relativistic spacecraft, for it would be unable to receive radio communications—these being restricted to the velocity of light. Presumably, if faster-than-light travel is possible, then some kind of tachyon communication could also be devised. However, this possibility raises real logical paradoxes. Under certain circumstances it should be possible to find out that an event has occurred before (in your frame of reference) it has done so, and then to take steps to prevent its happening. Life is already sufficiently complicated without admitting such possibilities.

At this time I cannot see any way in which faster-than-light travel *through space* can be achieved in practice, but I would hesitate to close the door completely. It may be forever unattainable, both in theory and practice, but at least the discussion of tachyons suggests there may be potential chinks in the apparently insuperable light barrier.

Bending the Rules

Let us accept for the moment that we cannot travel through space at or faster than light speed. Perhaps we can get from place to place without travelling through the intervening space, and perhaps in no time at all. Possibly we can even make a return journey, getting back before we set out! It sounds very much like science fiction of the space opera variety. However, the past couple of decades has seen a great resurgence of theoretical and experimental research into gravitation theory—and General Relativity in particular—which has led to a fuller understanding of the bizarre effects which may occur in extreme gravitational fields, and to the serious discussion of such enigmatic objects as black holes, white holes and wormholes. Could these provide 'gateways' between different regions of space and time which would allow us to bypass the light barrier and so fully explore the cosmos? To try to place these ideas in context, we must first look more closely at what lies behind the black hole concept.

The Black Hole

A black hole is a region of space into which matter has fallen and from which *nothing* can escape; the gravitational field inside a black hole is so strong that not even light can get out. In a sense, the idea of a black hole dates back to a suggestion made in 1798 by the great French mathematician Pierre de Laplace that there might exist massive bodies which were invisible because the escape velocities at their surfaces were greater than the speed of light. We have frequently talked about the idea of escape velocity in this book, and we know that the value of escape velocity at the surface of a body depends on two quantities, its mass and its radius. For example, a body with the same radius but four times the mass of the Earth would have an escape velocity of twice the Earth's value (i.e., of 22km/s). On the other hand, a body of the same mass but four times the Earth's radius would have an escape velocity equal to half the Earth's value. Laplace argued that if enough mass were compressed within a small enough radius then light would be unable to escape.

Laplace based his discussion on Newton's Theory of Gravitation, which regarded gravity as a force that acts directly between massive bodies. The modern view of black holes is based on a more sophisticated theory of gravity, Einstein's General Theory of Relativity. In Newton's theory, and in everyday experience, material objects have three dimensions—length, breadth and height—and time is something separate which flows past at a uniform rate no matter where you are in the Universe. Einstein's theory embodies the concept that the three dimensions of space and the dimension of time are intimately linked together making up what has come to be known as spacetime. Matter 'bends' spacetime in its vicinity, and it is this curvature of spacetime which gives rise to the phenomenon which we call 'gravity'. The greater the mass of material contained within a given volume, the greater the distortion of spacetime and the stronger the observed gravitational field.

In Newtonian theory, a planet moves round the Sun because it is influenced by a force (gravity) which acts over a distance and is directed towards the Sun. In Einstein's theory we say that a planet follows its natural path in the curved spacetime in which it is located; i.e., the planet is influenced by the curvature of spacetime in its *vicinity*, not by a 'force' exerted by a distant body. Admittedly this seems a rather abstruse way of looking at gravity, but it is not difficult to draw

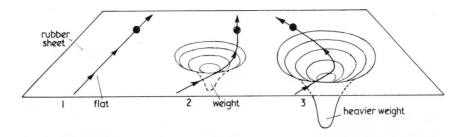

Fig 23 **The 'rubber sheet' analogy.** If we represent space as a rubber sheet which is flat in the absence of massive bodies, but is distorted when bodies are present, then we can see (track (1)) that a ball will roll along a straight line where 'space' is flat. A weight causes an indentation in the sheet, and a heavier weight a deeper indentation. Track (2) takes the ball through the outer part of the depression due to the first weight with the result that the ball's trajectory is curved. Track (3) is deflected to a greater extent by encountering the curvature of 'space' (i.e., the sheet) associated with the larger weight.

an analogy. Imagine a large flat rubber sheet; if it is perfectly flat then a billiard ball will roll across it at uniform speed in a straight line (analogous to Newton's first law in 'flat' spacetime in the absence of any matter). A weight placed in the sheet will cause an indentation (fig. 23) and the billiard ball will be deflected by the curvature of the rubber sheet. A heavier weight (corresponding to a more massive body) will cause a deeper indentation in the sheet (just as a more massive body gives rise to a greater curvature of spacetime, and a stronger gravitational field is experienced by objects in its vicinity). Of course this is only an analogy using a two-dimensional rubber sheet, whereas we have to think of spacetime as being curved in four dimensions—but it is, I hope, a helpful one.

Newton's theory is simpler, and perfectly adequate for all practical purposes in everyday experience. It is only under fairly extreme circumstances that General Relativity is superior. For example, in Einstein's theory, although a ray of light is the fastest-moving entity (it follows the 'straightest possible' line in curved spacetime) it is deflected when passing a massive object. This effect can be detected in rays of light passing close to the edge of the Sun. Although the effect is tiny it can be observed, and such observations constituted one of the first tests of the validity of Einstein's theory. In dealing with black holes, where we are treating very strong gravitational fields, we have to employ General Relativity and consider curved spacetime.

In 1916, using this theory, Karl Schwarzschild showed that if a certain amount of matter were enclosed within a sufficiently small volume then spacetime would be so severely curved that it would form a closed region out of which nothing could escape. The formula he obtained for the size of this region is a very simple one and, oddly enough, is just the same as the one which Laplace had obtained, for quite the wrong reasons, over a century earlier. The radius of this region is called the Schwarzschild radius,* and the region of space out of which nothing can escape is a black hole. A black hole should be very black indeed; it would emit no light and, no matter how much light you shone upon it, nothing would be reflected back. It is most important to realize that the boundary of a black hole is not solid. Once it has formed, the volume of the *matter* inside has no bearing on the size of the black hole—the radius of the hole depends only on the *amount* of matter it contains, and is equal to the Schwarzschild radius no matter how far the matter is compressed.

The boundary of a black hole is called the event horizon for the very good reason that no knowledge of events which occur inside the boun-

---

*Where R_s denotes the Schwarzschild radius, G the gravitational constant, c the speed of light, and M the mass contained in the black hole, the formula is

$$R_s = \frac{2GM}{c^2}.$$

dary can ever be communicated to the outside world. A ray of light emitted radially at the event horizon would stand still forever; any light emitted inside would fall towards the centre.

To give some idea of scale: if—in a fit of megalomania—someone decided to make the Earth into a black hole, he would have to compress the Earth into a radius of less than *one centimetre*. The Sun, over three hundred thousand times the Earth's mass, would have to be packed inside a radius of about three kilometres before light could not escape from its surface, and even a massive star, fifty times the mass of the Sun, would form a black hole only 150 kilometres in radius. Our entire Galaxy of one hundred thousand million suns could be contained within a black hole 0.01 parsecs in radius!

Do such highly compressed objects exist, and if so how are they formed? One way in which black holes may be generated is from massive dying stars. Normal stars like the Sun ('main-sequence' stars) shine and support themselves by means of thermonuclear reactions going on inside their hot dense cores. The energy so produced sustains the outward-acting pressure that resists a star's gravity, which otherwise would compress the star, and for so long as a star can sustain its energy output it will remain inflated like a balloon. But all stars have finite reserves of fuel and there must come a time when these reserves are exhausted. When this happens, a star can no longer support its own weight, and must contract. The majority of stars will spend the final part of their lives as highly compressed white dwarf stars which are comparable in size with the Earth and so dense that a teaspoonful of their matter, if brought to the Earth, would weigh several tonnes.

More massive stars go through their life cycles more rapidly and may end their lives by exploding violently as supernovae, much of their material being scattered into space. The remaining core of such a star is likely to be compressed to such an extent that it forms a neutron star, about ten kilometres in radius and so dense that a teaspoonful of its material would weigh thousands of millions of tonnes on Earth. We can see plenty of white dwarfs, and observations of objects known as pulsars indicate that neutrons stars, too, can be detected.

Theory suggests that any dying star whose final mass, after all possible means of shedding excess mass have been exhausted, is greater than about two or three times that of the Sun cannot end up either as a white dwarf or as a neutron star; instead it will collapse indefinitely under its own weight. Once the collapse has begun no known force can halt it, and in principle it will continue until all the material of the collapsing star is compressed to an infinite extent, giving rise to what is known as a singularity. Before this state of affairs comes about the star will have passed within its own Schwarzschild radius, and a black hole will have formed.

There are plenty of stars around with masses greater than this limit. Although many will become supernovae, a fair proportion seem destined to collapse into black holes. Since these massive stars go through their life cycles in tens of millions of years, and the Galaxy is over ten thousand million years old, there has been plenty of time for quite large numbers of black holes to form.

Detecting black holes obviously presents problems, but these are not necessarily so severe as might first be imagined. Although they emit no light or radiation from within their boundaries, they still exert a *gravitational* effect on their surroundings. If a black hole were a member of a binary system (two stars revolving around each other) then we might be able to detect the motion of the visible star due to the invisible companion. Furthermore, any gas falling in towards the event horizon will be severely heated and will emit X-rays (once it crosses the boundary, of course, it can no longer be detected). A very good case of a binary with a massive invisible companion which is also a strong compact X-ray source is an object called Cygnus X-1 which, located at a distance of some 6000 light years, seems the best candidate so far for containing a black hole. However, there is as yet no absolute proof that black holes exist and we must treat this 'discovery' with caution.

Any material object approaching close to a black hole has very little chance of escaping, for to do so it would need to be able to accelerate rapidly to velocities near that of light. The closer a spacecraft approached the black hole the less chance it would have of avoiding being sucked in, and once it reached the event horizon it could not escape at all. Any spacecraft which entered a non-rotating black hole would fall to the centre in a very short time and would be crushed out of existence by the infinite gravitational forces there. A spacecraft would fall from the event horizon to the centre of a 10-solar-mass black hole in about a ten thousandth of a second; for the crew the end would be mercifully swift! Even before the spacecraft reached the event horizon it would be torn apart by gravitational tidal forces—the parts of the spacecraft closer to the black hole would be subject to a much stronger attraction than the more distant parts of the craft, and the difference would be more than enough to tear any spacecraft to shreds.

Navigational Hazards?
On the face of it, black holes seem to be objects to be avoided like the plague, hazards to interstellar navigation rather than assets to the space traveller. Although we might deduce the presence of a black hole from its effects on a neighbouring star or from the radiation emitted by infalling matter, an isolated black hole in space would be a very different affair—a tiny black object against a black background which would be much harder to find than the proverbial black cat in a coal cellar. True, there is a very slim chance that if a nearby black hole were interposed between the spacecraft and a bright background object then light passing the hole would be distorted (in effect the black hole would act like a lens) and the black hole's presence might be revealed. Con-

ventional means of detecting obstructions, such as radar, certainly would not help, for the radar beam simply would be swallowed up and the operator, receiving no reflected signal, would conclude that nothing lay ahead.

One way of detecting a black hole in the immediate vicinity would be by sensing the gravitational tidal effects. With a 10-solar-mass black hole tidal effects should become easily measurable at a range of about 100,000 kilometres, but a vehicle heading for a black hole at, say, one third of the speed of light would enter the event horizon within one second of sensing the tidal forces. Perhaps a better approach would be to sense the unexpected acceleration of the spacecraft by the Doppler shift in the frequency of a communication beam from Earth, but even so a spacecraft is unlikely to be able to do much to avoid falling in.

Despite my painting a gruesome picture of the hazards of encountering a black hole, the fact remains that they would pose an utterly negligible risk to interstellar voyagers. Firstly, they must be very widely separated from each other; the mean distance between stars is something like a parsec (some thirty million million kilometres) and black holes are likely to form from only a small proportion of stars, so that they will

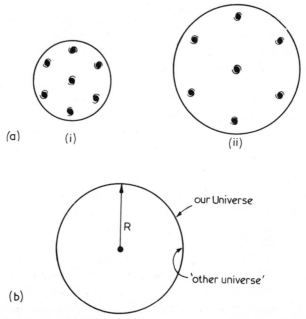

Fig 24 **Expanding universes.** If we represent our Universe by the outside surface of a balloon (the galaxies being represented as shown), then between (a) (i) and (a) (ii) the 'Universe' has expanded and each galaxy has moved away from every other galaxy; no one galaxy can claim to be the centre of this expansion. In (b), if our Universe were the outside surface of the balloon we could imagine another universe to exist on the inside surface, a universe of which we are not aware. As the radius (R) of the balloon increases, so both 'universes' will expand.

be much further separated. Secondly, they are very tiny (stellar mass black holes must be less than about 200 kilometres in radius). The chances of getting dangerously close to one must be vanishingly small. If a spacecraft were ever to enter a black hole it would be the result of a freak never-to-be-repeated chance event—or the result of a deliberate expedition.

Spinning Black Holes

Are black holes to be avoided at all costs, or is there more to the story than this? Symmetry seems to be a common feature of the Universe, and so it has been argued that if black holes represent regions where matter effectively disappears from our Universe there may also exist white holes from which matter suddenly begins to pour *into* our Universe. A white hole can be regarded as a time-reversed black hole—i.e., as a black hole running backwards!—and there is no objection in principle to such an idea within the framework of General Relativity. It has been argued by some that matter which falls down a black hole reappears out of a white hole somewhere else in the Universe. There is no proof that white holes exist, but the speculation is an interesting one. However, if a spacecraft is crushed to destruction in a black hole before being spat out of a white hole, surely such a method of transportation can be at best an academic possibility?

The General Relativity equations describing spacetime in the vicinity of a black hole have a symmetry which suggests that there exists 'another universe' on 'the other side of' the black hole. The curvature of spacetime increases as the event horizon is approached and then decreases into the 'other universe'. We cannot conceive what this other universe might be, nor can we be sure the mathematical solution has any physical reality, but it does appear as if black holes may provide the link between two quite separate universes.* It has also been argued that the spacetime of the Universe may be so arranged that the 'other side' of the black hole corresponds not to a separate universe, but to a different region of spacetime in our own. Perhaps matter falling into a black hole could pass through a sort of 'throat' and re-emerge in another part of space and time. Such hypothetical linkages between different regions of spacetime have come to be known as 'wormholes'.

---

*It is impossible to visualise another universe coexisting with our own, but the following may help. If there is sufficient matter in our Universe it may be so curved as to form a closed system, finite yet—so far as we can determine—unbounded. We can represent our Universe by the two dimensions of the outside surface of a balloon (fig. 24), and represent galaxies by spots on the balloon; we have to imagine that the space outside and inside the balloon is quite unknown to us and is inaccessible—we can move only on the balloon's *surface*. The expansion of the Universe can be regarded as the expansion of the balloon. As the balloon gets larger, so the spots (galaxies) move away from each other. If we can visualise our Universe in this way, then we can visualise another separate universe on the *inside* surface of the balloon. We have no contact with it except through black and white holes. Perhaps matter from our Universe disappears through black holes to appear in the 'other universe' out of white holes. This other universe, incidentally, would expand as does our one.

Unfortunately we cannot expect to be able to take a spacecraft through the wormhole associated with a non-rotating black hole. Quite apart from the nasty effects of tidal forces, the only possible route which a spacecraft can follow inside the event horizon takes it to the central singularity where it is crushed to destruction.

I have been careful to stress the phrase 'stationary non-rotating black hole' in the foregoing because the spacetime associated with a spinning black hole is markedly different. Real stars rotate and, if black holes are generated by the collapse of massive stars, then we would expect them to be rotating, too, and at a rapid rate. While the spacetime of a stationary black hole appears to be connected to another universe, the spacetime of a spinning black hole appears to be connected to an infinite number of 'other universes' and, most importantly, it does not necessarily follow that a spacecraft which crosses the event horizon *must* fall into the singularity. Some routes would lead to the singularity resulting in the destruction of the starship, but *without exceeding the velocity of light* at any time, the spacecraft could follow a trajectory which avoided the singularity and emerged from the spinning black hole into another universe—it could emerge in any one of uncountable numbers of 'universes'. This seems rather hard to comprehend, but if the space-craft falls in and does not get crushed up, yet cannot re-emerge from the black hole into our own Universe, it must end up somewhere else!

Sadly the crew would not be able to return to their own space and time by re-entering the black hole, for in order to do so they would have to exceed the velocity of light. A further possibility is that the set of universes is so joined that it may be possible to enter another spinning black hole and re-emerge in our own Universe. In effect, then, spinning black holes could provide the 'wormholes' connecting different regions of spacetime in our own Universe. If this is so it should be possible, by charting a suitable route, to travel from place to place effectively in no time at all. Spinning black holes may provide links between different regions of spacetime such that in principle it should be possible to emerge almost anywhere *at any time*, past, present, or future. Let me stress that these ideas are entirely hypothetical—although they do seem to be theoretically possible—and should be treated with great caution.

Even allowing that spinning black holes can connect regions in our own Universe, there would be many problems facing the interstellar navigator. For example, there must be a great deal of uncertainty in such voyages—the slightest error in trajectory within the black hole could place the traveller somewhere totally different from the desired location in spacetime, probably with no means of finding a way back. Indeed, without the benefit of the experience of preliminary expeditions, it is difficult to see how the entering astronaut could have any idea of his likely destination. Perhaps a future generation will find a way around such problems.

A more obvious and serious problem would be the tidal forces encountered by the starship and its crew as they approached the event horizon of the chosen black hole. It is worth looking in more detail at these forces. If you stand upright on the Earth's surface then, because your feet are closer to the Earth's centre than is your head, they are subject to a slightly stronger attraction than your head. The difference, normally, is negligible, amounting to a net force between your head and feet of a few millionths of your weight. In the neighbourhood of the event horizon of a black hole containing, say, ten solar masses, the forces would be colossal, something like ten million million times greater than the effect we experience here on Earth; we should find ourselves being torn apart on a cosmic rack by a force some ten million times greater than our normal *body weight*. For how long can you support yourself hanging by your arms from a beam? If you think you can manage a few minutes, imagine having someone else hanging onto your feet; then imagine having the entire population of London or New York hanging round your ankles! That is the magnitude of the tidal force to which we should be subject—albeit briefly—near the event horizon of a black hole, spinning or otherwise.

A spacecraft approaching such a black hole would be torn to shreds, and likewise the crew, before even crossing the event horizon; the fact that their constituent atoms might end up in another region of spacetime would be of very little consequence to them. On the other hand, the tidal forces near a black hole depend *inversely* on the square of the black hole's mass; the more massive (and hence larger) the black hole the weaker the tidal forces near its event horizon. For example, the tidal stresses at the event horizon of a 100-solar-mass black hole are only 1% of those encountered at a 10-solar-mass black hole, while a black hole of one hundred million solar masses would give rise to tidal forces at its event horizon less than those which we normally experience. If we set as the limit of tidal stresses which we would wish to endure on a space-flight a force equal to body weight, then we dare not enter a black hole less massive than about 30,000 solar masses. Given that some kind of protection could be supplied, and that the stresses would last only briefly, perhaps the human body could withstand tidal forces of 1000 times body weight. Even so a black hole of about 1000 solar masses would be required. To have any hope of surviving the trip, future space travellers are going to have to utilize very massive black holes.

Herein lies a problem. The most massive stars known are less than 100 solar masses and so cannot provide the right kind of black hole. However, there is no theoretical reason why much more massive black holes (formed by the collapse of large masses of material) should not exist. It has been speculated that massive black holes may exist in the nuclei of galaxies, and that our own Galaxy may contain a black hole of some hundred million solar masses. Possibly, too, the huge globular star clusters—of which our galaxy has more than a hundred—may contain massive black holes.

So far as we can tell at the moment it is most unlikely that any really massive black holes exist near at hand; i.e., within even a few *thousand* parsecs. Clearly black holes are unlikely to provide a short-cut route to local interstellar travel, but it may be that massive black holes in galactic nuclei provide the key to intergalactic travel for those societies which have already mastered the art of interstellar travel at relativistic velocities. Massive spinning black holes may also provide a means of circumventing the famous 'twins paradox'. By choosing to re-enter our Universe at an appropriate point in *past* time, and then returning to Earth at near-light speed, the ageing of the astronaut could be exactly matched to that of his Earth-bound twin.

We have stretched the rules, and our credibility, rather too far already but—bearing in mind that we are dealing with speculation—let us imagine a possible black hole mission. The crew of the *Final Frontier* are despatched to investigate a distant galaxy. Using a suitable propulsion system, such as the interstellar ramjet, which picks up its fuel as it goes along, they are able to fly to the galactic nucleus at a steady 1g acceleration. Although the crew feel themselves subject to a constant acceleration (and therefore feel their normal weight), as seen from the Earth their speed builds up rapidly at first and approaches closer and closer to the speed of light without actually getting to it (Table 4 relates speed to distance travelled to time dilation for 1g acceleration).

Because of the increasing time dilation effect, the astronauts find they can reach the black hole at the galactic nucleus in about *ten* years, although—because the galactic centre lies at a distance of 10,000 parsecs—30,000 years will have elapsed on Earth. Entering the spinning black hole (assuming for the moment that it exists) on their pre-computed trajectory, they feel no severe tidal stresses. Passing through the bizarre world of the interior of the black hole they re-enter our Universe in the heart of the target galaxy. After five years' exploration they enter another massive black hole to return to our Galaxy.

Let us suppose that they re-emerge 10,000 parsecs away from the Earth. According to elapsed time on Earth they have been away for rather more than 30,000 years, although the crew have aged only by some 15 years. In order to meet up again with their contemporaries on Earth they have chosen a route in route in spacetime so as to appear about 30,000 years in the *past*. By the time they reach the Earth—after travelling 10,000 parsecs since their re-emergence—if they have got their sums right, they should land on Earth 25 years later than the date on which they first set out. Their earthly companions will be 25 years older than they were on the launch date and the astronauts themselves will have aged by 25 years (10 plus 5 plus 10).*

---

*The cunning astronaut may chose to plan his trip rather differently. Before departure he arranges for his salary to be invested at a good rate of interest (and his salary is to be paid while he is away). By making an appearance 60,000 years in the future he would collect a handsome sum. Taking off again he plans a route that takes him back to his own time.

Is this conjectural scenario a real possibility? There are so many 'ifs' in the chain of argument that even my own credulity (which is considerable) is strained past breaking point. The logical contradictions and paradoxes which would flow from such a possibility are legion. For example, the astronauts would have returned while their spacecraft was still (theoretically) visible through a powerful telescope to be continuing its outward journey. In essence they would have become time travellers and the paradoxes associated with that possibility have been well explored in science fiction. I am inclined to think that journeys of this type are likely to be confined to the pages of science fiction but, again, I would hesitate to dismiss out of hand even so far-fetched a possibility. At least the idea of black hole transportation offers up another possible crack in the light barrier.

I should like to conclude this black hole frolic by making an appeal to any future reader who lives in an era when the problems of spacetime travel have been overcome—if such an era should ever arrive. Please make a trip back to 1978 and let me know (*prior* to publication) if the scheme really works. The twentieth-century reader may draw his own conclusions regarding this request from the way that this paragraph appears in print.

Other Ways Around the Rules

Many other possibilities have been raised—'thought travel', 'teleportation' and the like—but I do not feel competent to comment on these. A favourite theme in space fiction television series is the idea of transporting men and materials by means of some kind of beam ('Beam up three, Mr Scott'). Perhaps the necessary information to reconstruct a copy of an individual could be communicated by electromagnetic radiation to a distant colony possessing the technology to carry out this operation, but that would scarcely be the same as transporting the individual himself. Alternatively, if the individual could be broken down into wavepackets corresponding to his individual atoms and beamed across interstellar space in this way then at least two factors would conspire against the reconstruction of the unfortunate individual. Firstly there is the 'uncertainty principle' which implies that we cannot fully specify everything about a particle; and secondly there is the effect of the interstellar medium on the beam to consider. There could be no guarantee that the exact same person would be reconstituted in every facet at the other end.

Speculation can continue without limit, and I have already taken it too far. The important point is that interstellar travel looks to be possible even without inventing wondrous new technologies. Given the appropriate development of existing technology the means to reach the stars could be available next century. Ways of breaking the rules would come as an added bonus, but breaking them is not essential.

PART III

Love and Hate

9 On Board Ship

How will life in the starships affect the travellers? What sort of living conditions, social and command structures will arise? Is it really likely that—as in the Starship *Enterprise*—we shall see a starship organised like a naval warship? In trying to answer these questions we have at least three different situations to consider:

(a) the flight time is many times longer than a human lifespan—the case of the Space Ark;

(b) the journey may be completed within a single human lifetime, but occupies a large fraction of that time—the situation which would arise with a 'fast' starship (velocity in the range $0.1c$ to $0.5c$), or with increased human longevity;

(c) relativistic spaceflight (velocity close to the speed of light) whereby even long interstellar flights may be accomplished within a small part of a traveller's lifetime.

Each of these possibilities has its merits and demerits, and raises its own characteristic problems. In what follows, I have set out for each of these situations what seems plausible to me; the reality undoubtedly will be quite different.

The Space Ark

In essence the idea of the Space Ark is that an entire colony of human beings, in a closed and self-contained environment, be despatched on an interstellar journey likely to last for hundreds, if not thousands, of years. The original crew—if 'crew' is a word that can be applied in this context —would set out in the full and certain knowledge that none of them would reach the target star; that privilege would fall to their remote descendants. Generation after generation would be born, live, work, grow old and die within the confines of the self-propelled colony. In effect, the starship would be a miniature planet. Such an idea is by no means new, and dates back at least as far as the writings of Professor J. D. Bernal in the nineteen-twenties.

To be workable, the Ark would have to contain a sizeable population providing a wide cross-section of skills and abilities to cater for all aspects of living: food production, catering, medical care, education and entertainment; and various forms of employment appropriate to a Space Ark: manufacturing industry, engineering (maintenance and control of on-board systems) and science (particularly astronomy and astrometry). There would also be a rôle for the spacefaring funeral director! The population would have to contain a sufficiently great genetic pool to ensure a supply of persons with the appropriate qualities of leadership to cope with the trauma of the ultimate arrival at their pre-ordained target.

The Ark would have to be self-contained in every way, and capable of coping with the hazards of interstellar space—radiation, collisions and the like. The space colonies discussed by O'Neill and others naturally relied on solar energy for their power supplies. The interstellar travelling colony would require an internal power source capable of sustaining the inhabitants for many centuries—the energy source being nuclear, thermonuclear, or some alternative (e.g., matter-antimatter annihilation). Multiple redundancy of essential components would cope with the inevitable failures, but a considerable industrial facility would be vital both to maintain the Ark and to handle the initial construction activity at the target system. The size of the population, the extent of the life support system (which cannot be replenished *en route*), the social amenities, industrial capacity, guidance and control systems, not to mention the propulsion unit, taken together indicate that the interstellar Ark must be a pretty massive structure.

Without considerable psychological and sociological research one cannot say for sure what would be the minimum population necessary to maintain a balanced community on such a vessel, but one thousand would probably be too few, and ten thousand is probably nearer the mark. Perhaps there will be developed techniques of mind control which will condition people to enduring a much more cramped environment for prolonged periods. Probably the closest approach to closed colonies in present-day experience exists in nuclear submarines which may remain at sea, submerged, for many months without any physical contact with shore bases. The habitable volume of a typical ballistic missile nuclear submarine is some 10,000 cubic metres and this is occupied by about 150 men; this works out at some 70 cubic metres per man.

By way of comparison, *Skylab*, the US orbiting laboratory of 1973/74 had a habitable volume of 292 cubic metres and a mass of about 90 tonnes (nearly 100 cubic metres and 30 tonnes per man), and in this environment a crew of three was able to function quite happily (apparently) for up to 84 days. The habitable volume in the Soviet *Salyut* orbiting space laboratory is about 100 cubic metres and in the case of Salyut 6 (1977/78) a crew of two endured 96 days and even found room to entertain visitors for a few days!

Little can be deduced from these cases as, with skylabs and nuclear submarines, one can return to base in a quite short time if the need or desire arises, and there is the possibility of help being rendered in case of difficulty. The interstellar colony could hope for no help from the Earth —in any case, if some emergency arose after it had travelled even a few light-months from the Sun, by the time the message had reached Earth, and help had been despatched even in the form of a near-light-speed 'service vehicle', the assistance it might be able to render would probably be too late anyway.

Unless the Space Ark were being despatched as a desperate measure to preserve only a few members of the human race in the face of imminent catastrophe, the colonists would wish to have the human resources to cope with all likely eventualities, and to have living conditions which were of the highest standard. If by some form of mind control successive generations could be conditioned to endure a cramped and limited environment, what would be the point of such an undertaking? There would be no point at all in unloading a package of neo-human zombies (who would probably suffer from acute agoraphobia on disembarking) in a distant planetary system!

If we take a population of 10,000 as a working figure, and if we are to provide them with reasonable living conditions, then we are thinking in terms of a self-propelled version of an Island One type of colony, the mass of which has been estimated as being 3.6 million tonnes. After adding extra mass for the propulsion system and additional shielding, the mass of the Ark will exceed 4 million tonnes. The problem of accelerating such a starship to any reasonable sort of speed is rather formidable.

Assuming some kind of fusion drive or nuclear pulse rocket system, and taking as a working value an exhaust velocity of 10,000km/sec, then a mass ratio of 2.72 would yield a final spacecraft velocity of 10,000km/sec, or about 3% of the speed of light. Ignoring the mass of propellant tanks and the like, something like 7 million tonnes of fuel would be required just to accelerate the Ark to its cruise velocity. And presumably the inhabitants, or their descendants at least, would wish to stop the spacecraft when they reached the target. The mass ratio for the deceleration phase would again be 2.72, making a mass ratio for the entire operation of 2.72×2.72; that is about 7.4 (i.e., about 26 million tonnes of fuel, excluding tank mass, etc.). Although this ratio could be eased by using a ramscoop 'parachute' or an electrostatic drag sail, the inevitable conclusion is that self-propelled colonies of this kind must be—for the foreseeable future—restricted to speeds of a very few percent of light speed.*

*G. L. Matloff (*Journal of the British Interplanetary Society*, vol 29, pp 775–785, 1976) has calculated that a 'Model I', cylindrical O'Neill colony for 10,000 inhabitants could be accelerated to a cruise velocity of 0.01c by the Daedalus probe motor, assuming 150,000 tonnes of fuel to be available (the current Daedalus proposal uses only 50,000 tonnes). He also estimates that about 600 tonnes of fuel would allow a fusion reactor to supply adequate on-board electrical power for a 1000-year journey. It is interesting to note that the technology to build colonies should soon be available, that Daedalus technology is not too far in advance of present-day technology, and that working fusion reactors, hopefully, will be with us by early next century. The building of a Space Ark capable of reaching Alpha Centauri in about 400 years of travel time should be within our capabilities by the middle of the twenty-first century.

If we take a cruise velocity of $0.02c$ as a reasonable practical value, then a three-parsec journey will occupy 500 years and many generations will pass before a landing is achieved. If we assume for the moment an average lifespan of 75 years, with couples having two children at about age 25, then nineteen clear generations of children will be born on the flight. One essential feature of the Ark ecosystem will be zero population growth, the replacement rate precisely equalling the death rate. The slightest imbalance would be fatal for a closed isolated system such as this. For example, if—on average—the number of children born to each couple were 2.01 instead of 2—i.e., if one couple in every hundred had three children instead of two—the population of the Ark would increase twelve-fold in the 500-year flight!

How would life in the Ark affect the family unit? It might be that the circumstances would favour a commune or Kibbutz-type structure, with the caring for children attended to communally or by a central agency. Likewise it is possible that the colony situation would favour an open and permissive society. I am inclined to doubt this. In the absence of good evidence to the contrary it seems more likely that—as in many of the great human treks of the past—there would be a strengthening of the marital and family bond, with marital relations being tighter rather than looser. With its long tradition behind it, the family unit should make for a more stable kind of society in the Ark than would the more fluid open society. Initially, at least, this would be most important. Another factor which might affect whether or not a traditional style of society was established in the colony would be the age structure of the original inhabitants of the Ark. An initial community made up entirely of young families might well adopt a different social structure to a community containing the full spectrum of age groups.

The community would be much more akin to a city state such as Athens or Sparta in ancient Greece (or the flying island of Laputa in *Gulliver's Travels*) than to the conventional idea of a spaceship. The idea of having a captain, officers, crew and passengers in this situation is clearly absurd. Instead, one would have some kind of civilian administration. Whether this would be authoritarian or democratic can only be a matter of conjecture. With so many aspects of the life of the community predetermined in any case, the only areas over which the administration would have meaningful control would be social activities and priorities, education, and the like. A society whose material needs are fully catered for, and whose material condition cannot be improved, might well evolve a social order governed by a mass of regulations designed to give a sense of power and purpose to the upper echelons of the hierarchy. We simply have not the experience to judge what might happen to a community which remained closed for five hundred years; all we can say is that it will evolve in its own way, and that the rest of humanity, and the Earth in particular, will assume less and less importance in the minds of the travellers, despite the availability of means to communicate with Earth. Once the turn-round time for a message exceeds a human lifespan, interstellar conversation becomes a little pointless!

The practical, economic, emotional, social and moral problems associated with this approach to interstellar travel are of the highest magnitude. Life in the Ark may become a most pleasant form of existence. On the other hand it may become a living Hell from which there is no escape. If we assume that all economic and technical problems can be overcome we are still faced with a multitude of enigmas, some of which are discussed below.

Who today would undertake a mission whose target was unattainable and from which no return was possible? 'Suicide missions' have been undertaken in wartime conditions, missions from which the chance of a safe return was practically zero. The outcome was often as predicted, but people have survived in the face of 'impossible odds' and in such situations there remained some ray of hope. With the interstellar Ark there would be no hope of a return—the severing of relations with the community of Earth would be absolute. Granted that sufficient people were to volunteer, to be trained, coerced or press-ganged into undertaking this venture, would they or the mission planners have the right to make a decision which committed countless generations yet unborn to pursuing their objectives. In a sense, of course, this happens all the time. The results of our deeds and misdeeds on this planet come home to roost for future generations; the crises we face on this planet today are— in part at least—due to the decisions or indecisions of our forefathers. But the options open to the individual on Earth are much wider than those open to the inhabitants of the Ark. Likewise, the inhabitant of an L-5 colony, unhappy with his lot, could always choose to leave that colony. The only escape from the Ark would be to take a (fatal) walk outside.

The problem hinges on the age-old question of the value, rights and needs of the individual, and whether or not these should be subjugated to the greater good of The Cause, whether that be the state or the species. In western nations the freedom of the individual and the rights of the individual have for long been regarded (in principle at least) as being of paramount importance; other cultures do not rate the individual so highly. It may well be that a more highly developed society will find it quite acceptable to make long-term decisions on behalf of unborn generations. Perhaps those of this view will gather together on one colony, evolving a distinctive social structure, and it is they who will launch the first Space Ark.

Once the mission is under way, will there be any guarantee that commitment to *the mission* will be sustained for centuries? Will succeeding generations evolve quite different ideas as to their rôle and purpose? Will the community be able to sustain its scientific and technological expertise or its pioneering spirit for a period of time which may be as long as that which separates us from Christopher Columbus, or even

King Canute? Or is it inevitable that—assuming no disaster befalls the Ark itself—by the time the target is reached no one aboard will know or care that this has happened?

A strong sense of purpose would be vital to the community, and the goals of the society would have to be clearly defined and obvious to all. As the centuries rolled by, *the mission* might acquire a religious intensity. The resultant sense of being 'a chosen people' need not be dispelled by communications from Earth, which would relate to a bygone era on a distant world of which no one on board had direct or even second-hand experience. In the absence of an acceptable *raison d'être*, the colonists might well be overwhelmed by a feeling of utter futility leading to tension and strife, or just sheer apathy. Either way, the future of the Ark community would be bleak.

Further speculation at this point would be idle, but I cannot help feeling that the revolution in the human make-up required to make possible *successful* Ark missions may turn out to be greater than the technological hurdles which lie in the way of relativistic spaceflight!

A paradox facing intending crews of Space Arks concerns future technological development. As we have seen, the speed of transportation has escalated dramatically during the twentieth century; in the five decades between Kitty Hawk and Sputnik 1 the increase was by a factor of a thousand. Of course this increase is not linear; it proceeds in fits and starts with succeeding technological 'breakthroughs'. Nevertheless we may see an increase by a further factor of ten before this century is out. Would it be wise to undertake a 'terminal journey' when by waiting a while some new development might render the voyage attainable within one's lifespan? Why undertake a 1000-year mission when, in a hundred years' time, it might be possible to complete the entire voyage in 100 or even 10 years? It would surely be galling in the extreme for the remote dedicated descendants of the original pioneers who had sacrificed their own lives (and those of many others) to the goal of reaching (say) Epsilon Eridani to arrive after half a millennium in transit to find a fully colonized world, with a new technology, awaiting them—together with a welcoming party comprised of civic dignitaries, historians and a few space scientists with an interest in archaic craft.

The Space Ark may come, it may turn out to be the means by which mankind populates the Galaxy, but I suspect that the first men actually to reach the stars will do so by means of a high-speed vehicle, and within their own lifespans, too. Whether they would have been the first to *set out* is quite another matter.

The Big Sleep

In an effort to circumvent the problems associated with having numerous generations live and die in the Ark, various writers have suggested that the crew be placed in some form of suspended animation for the duration of the flight, to be wakened a little time before arrival at their destination. If a crew could be preserved in this way, without mental or physical deterioration, for hundreds of years then they should waken as fit and well as they were at the start of the voyage. In effect they would be no older than when they set out, and should be able to live out the remainder of their lifespan at the normal pace.

The idea has many attractions, and not only for space travellers. If medical science is unable significantly to extend the duration of human life, this means that one is restricted to a finite number of waking hours. But the technique of suspended animation could allow an individual to extend his 'allotted span' indefinitely—spending most of the time in a suspended state, and waking for the odd few months at pre-selected times! The sum total of *conscious* existence might be no greater than normal, but for those with a morbid desire to experience the future, suspended animation would offer the means to do so. It might be a salutary experience for futurologists to be able to test their predictions in this way.

A patient suffering from a serious, and presently incurable, disease might choose to be placed in cold storage until a cure were developed. Prospective interstellar travellers might opt to be suspended until such time as a really fast relativistic starship is developed which will allow them to make interstellar journeys of short duration. This might be more acceptable than placing them in suspended animation aboard a *slow* spacecraft.

As with most 'good ideas', the realization would bring many problems, social, psychological, and economic. If storage space were limited, who would have the right to be stored? Would it be the highest bidder, or would some kind of worthy tribunal decide each case on its merits? One could even envisage the *reductio ad absurdum* situation where by far the majority of the human population was in a state of suspended animation and most of the waking remainder was employed in tending the suspension centres. The temptation not to bother waking the suspendees might be quite great, as indeed might be temptation to 'pull out the plug' (particularly if the suspension fee were paid in advance). The cost of running such centres might be quite high and, although a suspendee in the early part of his life might have paid into an investment fund to cover his period of storage, inflation, or some other vagary of economic life, might wipe out his resources. In that case he would have to be wakened and forced to earn his keep in a society which might be quite alien and in which his previous skills were of no value. The awakening Rip Van Winkle might well have unpleasant surprises in store.

Should suspended animation eventually prove practicable, I feel it is unlikely ever to be used on a large scale on Earth; indeed, there are a great many advanced medical techniques today whose application is severely restricted. But for specialized purposes such as interstellar travel the benefits are clear. The weight saving could be very considerable, with the need to maintain a habitable biosphere for several hundred years removed; and there might be no need to have an Ark on the scale of an O'Neill colony (although the suspendees would require facilities when they did wake up, and might need to use the facilities of the mobile biosphere for some time after reaching their target).

A crew of suspendees would have made their own decision to go (at least one hopes so; the possibility of interstellar Arks being used to send undesirables to penal colonies should not be ignored), and they would not be committing future generations unwillingly to continue with what, to the original crew, 'seemed a good idea at the time'. Assuming all the on-board systems continued to function for centuries, a normal environment would be reconstituted within the Ark in time for the crew to be reactivated.* The sleepers would awaken just as committed to their task of exploration and colonization as on the day they embarked; they would, nevertheless, be faced with the possibility that during their centuries of slumber, and as a result of technological developments undreamed of when they set out, terrestrial colonists travelling in a fast starship had beaten them to it.

Suspended Animation—the Practicalities and Prospects

At present there are two approaches to the problem of suspended animation—freezing and hibernation.

The science of cryobiology is concerned with biological activity at low, and abnormally low, temperatures, and with the effects on plants and animals whose temperatures are depressed below the freezing point of water. Freezing leads to the *complete cessation* of bodily functions. A frozen astronaut, in principle, would require no sustenance other than the maintenance of his low temperature. The heart of the starship would be racks of individual deep freezes. All being well the crew would be thawed out in prime condition as they approached their target.

In the present state of the art, freezing is not an attractive proposition; experiments involving the freezing of animals have not given grounds for optimism about freezing humans. Some ten years ago, experiments carried out on golden hamsters revealed that none survived more than a few hours after thawing if the temperature of the fluid in which they were immersed was lower than about $-10°C$ and, again, none recovered if the *deep* body temperature sank significantly below $-1°C$. None revived fully (i.e., without damage) if more than 50% of bodily water had turned to ice, and no full recovery was attained if the period of freezing lasted more than an hour or so. Since that time, so far as I know, there has not been significant progress in the length of time for which animals may be frozen.

The problem lies in the damage which is inflicted both by freezing and by thawing. Different parts of the body cool at different rates, and

*The parallels with events discussed in the Arthur C. Clarke novel, *Rendezvous with Rama*, are quite striking.

it is impossible (short, perhaps, of skewering the victim with countless heat pipes) to ensure that all parts reach the same temperature at the same time. The formation of ice crystals disrupts cells and bodily organs; anyone who has thawed out a frozen lettuce will have noticed how limp and unappetizing it has become due to the damage inflicted on its constituent cells. The resulting concentration of salts, and the effects even of one's own digestive juices, would likewise be harmful. The frozen astronaut, assuming he survived after thawing, would most likely suffer from chronic stomach ulcers as a result of the corrosive effects of the hydrochloric acid in his own digestive system.

Perhaps the brain presents the greatest problem. Other body cells regenerate themselves, the body operating a system of continuous maintenance resulting in most cells being replaced in the course of a year; brain cells do not. The number of active brain cells, or *neurones*, is determined in childhood and thereafter declines at an average rate of roughly 10,000 per day. Admittedly there are many more neurones than are actually employed at any one time, but the fact remains that the supply diminishes with age, and damaged cells cannot be replaced. In our present state of knowledge the effects of massive brain damage are irreversible. Damage to the brain must at all costs be avoided. There would be little point in freezing fit, intelligent and mentally alert astronauts here, only to thaw them out as vegetables at the end of the voyage. That the brain is easily damaged is well known; the circulation to the brain need only be halted for thirty seconds to produce irreversible effects.*

A decade ago the prospect was held out that before long we might have 'organ banks' in which hearts, livers, kidneys, etc., might be held indefinitely in cold storage ready for 'instant availability' transplant surgery. Although damage to cells can be avoided in some circumstances by the injection of protective agents such as dimethyl sulphide, the situation today is that, with the exception of bone, skin tissues and the like, we cannot successfully freeze and store individual bodily organs. We are a long way indeed from the successful freezing of human beings—if it should prove possible *ever* to do this. (One must remember, as well, that frozen human beings are brittle, and likely to shatter if dropped.)

I am not optimistic on this score, but it is worth noting that, on occasion, individuals have suffered considerable lowering of body temperature and survived. In the year of the first Sputnik there occurred a notable case of a woman who had lain in a drunken stupor outdoors in an air temperature of below $-20°C$ with the result that her body temperature dropped to $18°C$—halfway between normal and freezing point. She survived, although with severe frostbite. A number of similar cases might lead one to suspect that the consumption of copious

*At lower temperature, however, the situation improves. For example, at 10C° below normal bodily temperature, circulation may be halted for some eight minutes without damage.

quantities of alcohol may be the key to surviving the freezing process, and that one of the selection tests employed to pick out prospective cryonauts might involve the ability to down a bottle of Scotch within a predetermined period. More likely this is an example of spurious correlation whereby the alcohol did not contribute to the survival, but was the primary cause of the individual being frozen in the first place.

Hibernation

If freezing appears to lie beyond our grasp, what is the situation regarding hibernation—which is, after all, a familiar process in the animal world. Hibernation results in a *slowing down* of body functions. The heartbeat can drop to as low as two beats per minute in some species, while the breathing rate falls in proportion, the breathing itself becoming very shallow. The body temperature may drop quite close to zero, but metabolic processes continue sufficiently to maintain a body at between 0.5° and 3° above the environmental temperature; if the ambient temperature drops below zero, the metabolic rate increases and the animal may wake up. Indeed, during the hibernation season, mammals do waken repeatedly. The activity of the nervous system is reduced, but it can still respond to certain stimuli. The animal is not anaesthetized; an attempt, say, to carry out surgery on a hibernator would result in its rude and painful awakening.

One feature of hibernation of interest to mission planners and the designers of starships is the greatly reduced metabolic rate, which implies that hibernating astronauts would not require much feeding. In a human body cooled to 20C° below normal the metabolic rate is less than one quarter of the normal value, and in a state of true hibernation it would be much further diminished.

The mechanism of hibernation is not really understood; it amounts, in effect, to a kind of self-induced hypothermia. The ability to hibernate appears to be restricted to a small range of mammalian species, mostly rodents such as the dormouse, squirrel and hamster; the hedgehog, however, is an insectivore. Hibernation in these animals is preceded by a period during which a store of food is laid in or in which the animal gorges itself until bloated with food. Among closely related species there may be some which can hibernate and others which cannot. The ability, then, may be under genetic control.

The science of molecular biology has been the centre of considerable interest and controversy recently because of significant advances in the area of genetic engineering, the ability to tinker with the code of inheritance which governs the replication of living species. Heated argument has centred on the technique known as recombinant DNA technology, by means of which hybrid strands of DNA (the 'genetic code') can be concocted by sticking together pieces of DNA from different origins. The possible benefits to be derived from such a technique range from the synthesis of drugs to the possible repair of human genetic damage. The fear concerns the ability to produce by accident or design the 'doomsday bug' which, released into the atmosphere, might wipe out the great majority of the human race. This favourite theme of fiction may not be so far removed from reality. The development of recombinant DNA technology has led, perhaps for the first time, to serious and quite widely supported demands being made by *some workers in the field* that severe restrictions be placed on an area of research. There are those who would like to see a complete ban on genetic tinkering, but such a ban would scarcely be likely to be effective. Pandora's box, once open, cannot be closed.

Following the first recorded construction in the laboratory of a biologically active gene in August 1976 by a team based at the Massachusetts Institute of Technology there have been developments in synthesizing genes from scratch. Although the technical difficulty in such an approach is of a very high order it may well be that approaching genetic engineering by the insertion of wholly synthetic genes may prove to be safer than sticking together naturally occurring DNAs.

Either way, however, the door is now open for us to modify our own biology. In the past, the evolution of life on Earth was governed by naturally produced mutations and the process of natural selection; in the future we may be able to control our own evolution. At this stage one can only speculate on what might be possible. It may be that breeds of humanity can be produced which are ideally suited to environments hostile to ordinary humans, or perhaps the lifespan can be increased, so obviating one of the major difficulties in the way of interstellar travel. Or, again, it may be possible to produce humans who are natural hibernators.

Ageing is not halted by hibernation, but there is good evidence to suggest that bodily wear and tear are markedly reduced with a resultant extension of lifespan. As a general rule of thumb, lifespan is related to body size, and small hibernating mammals do seem to live longer than non-hibernators of the same size. By analogy with small hibernating bats and similar-sized mammals we might reasonably hope that, with the application of hibernation, the human lifespan could be extended by a factor of five to ten (although the period of *conscious* existence might not significantly be improved). A factor of ten would render moderate Ark journeys feasible; for example, a three-hundred-year hibernation period would be acceptable. An advantage of hibernation over freezing would be the relative ease with which crew members could be aroused from this state.*

On board ship it would probably be politic to maintain a small proportion of the crew on watch at any one time, adopting a rota of, say, one year awake for every thirty in hibernation. Given some serious alert,

*Hibernation is a totally different process from sleep (in which body temperature is normal and bodily functions proceed at their normal rate). 'Waking' from hibernation to a state of full awareness takes far longer than waking from sleep.

the entire crew could be mobilized quite quickly. I suppose that in the case of frozen astronauts repeated freezing and thawing might be possible, but the prospect of serious damage to organs would be even further enhanced by this process. Furthermore, failure of the freezer system would lead almost certainly to the death of the cryonaut, while a sharp environmental change should merely waken the hibernator. Altogether, hibernation seems a better prospect than freezing, at least for flights of moderate duration. Even on high-speed ships, with journeys of only a few years' duration (ship time), hibernation for a fair proportion of the journey would have much to commend it, as it saves on life-support resources and minimizes crew boredom.

The waking crew on a long interstellar journey would be occupied on a multitude of scientific tasks and maintenance activities. In particular, the steady rate of progress of the slow Ark would allow the complete mapping of the stars in the Galaxy, their parallaxes being determined over enormous baselines. But, with hibernation, no one need spend his entire life on these most valuable but mundane chores.

The Space Ark may take the form of a 'generation ship' with all the material necessities to cope with generation after generation of individuals who live out their lives in a mobile biosphere. Such a starship is not far beyond the limits of present technology, but does pose considerable social problems for its inhabitants. The biologists may yet provide the answer to the problems of long-term space voyages in the form of human hibernators, greater longevity or some other kind of genetic adaptation. The Ark may undertake only the single voyage to a particular stellar system which is to be colonized. Alternatively, it may be refuelled (from comets, or gas giants like Jupiter in the target system) and, after establishing a viable human colony, move on to another system. In this way a small fleet of Arks could spread humanity round the Galaxy at a leisurely pace, if this were deemed a desirable thing to do. Perhaps, though, the inhabitants of the Ark would grow so used to their way of life that planets would cease to have any appeal other than curiosity value, and they might settle for a nomadic existence, travelling endlessly from system to system, pausing only to replenish their resources.

The Fast Starship
Given that starships may eventually be constructed which are capable of travelling at a reasonable fraction of the speed of light—a tenth or a half, say—then a different approach to manned interstellar spaceflight would be possible. The nominal three-parsec mission which would have occupied the Ark for some 500 years would take 100 years to accomplish at a velocity of $0.1c$, or twenty years at $0.5c$. The former value is of the order of a human lifetime, and even twenty years is a substantial fraction. Let us accept the higher value of velocity for the moment. A three-parsec round trip would occupy at least forty years, perhaps

forty-five, including acceleration and deceleration times together with the exploration time at the target system, but it would certainly be possible for a healthy astronaut in his early twenties to undertake such a mission and return to Earth before his seventieth birthday.

He would not be moving fast enough for relativistic effects to have a really significant effect on his ageing process; over the entire journey he would 'save' about five years compared to his Earth-bound compatriots (i.e., the voyage would last for 45 years of Earth time and 40 years of ship time)*. When he arrived home friends and relatives would seem to have aged marginally more than he himself, but the effect would not be very obvious. The process of ageing varies considerably between individuals; we all, I am sure, know octogenarians who look and act as if they were sixty years of age, and sixty-year-olds who give the impression of being about ninety-five. Apart from the changes which had occurred on Earth over a period of forty to fifty years (which could be considerable) there would be no major problem of readjustment for the returning astronaut among his Earth-bound 'contemporaries'. Having spent his entire working life in space, he could look forward to a quiet period of retirement back home.

Forty years or so is nevertheless a very long time to spend confined in a starship (even a voyage to Proxima Centauri in a $0.5c$ ship would require a round-trip time of about 20 years) and a substantial biosphere and human complement seem essential to such a mission. People may still be born and perhaps die in the ship, so that we may be talking about a kind of 'mini-Ark'. If so, then we have a huge mass requirement and a colossal energy requirement to accelerate such a ship to half the speed of light. Could any degree of sanity be maintained after decades in space with only a small number of crew members? Possibly selection, training and motivation could produce the dedicated kind of crew needed to see through such a mission in limited living space seeing only the same few familiar faces year after year. But we all know how tensions build up and minor irritations assume grossly exaggerated proportions when a few of us are thrown together in cramped circumstances for even a few days. Serious and conscious effort has to be made by each individual to avoid conflicts. Despite these difficulties, though, I have no doubt that when such a mission becomes possible there will be no shortage of volunteers.

The command structure remains a problem. Certainly, one could have a captain and a chain of command, but the captain is unlikely to remain fit to command throughout the entire duration of the mission. For example, if an experienced astronaut aged forty is placed in command at the start of the mission, he would have passed his sixtieth birthday by the time that target was reached, and would be about eighty-five on his return to Earth. A system would have to be devised

*At the lower speed of $0.1c$ time dilation effects would be negligible over such a journey.

to ensure that a captain was appointed only for a limited period, and could be relieved of command if the situation warranted such a step. If reproduction were not possible on a starship of this kind, then the initial crew would have to contain young trainees in their teens who would reach the ranks of command by the time the starship was on its return flight. The selection procedure for commanding officers would probably have to be carefully controlled, although it is possible that a democratic election procedure might be successful (unlikely, but possible).

If the longevity of the crew, and in particular their 'prime' period, can be extended, then the problem would be much reduced. Hibernation, or some other kind of suspended-animation procedure, would offer the ideal solution in this sort of velocity range. As in the case of the Space Ark we might expect a few of the crew members to be awake at any one time, alternating on some kind of rota system with the 'sleeping majority' so that no individual need be conscious for more than a few years (or even months) of the flight time. The homecoming crew (if they do decide to come back) will still be able to meet old friends and relatives, but they would outlive them by a considerable amount after their return to Earth.

Given suspended animation, the automatic exploration 'scout ship' with a human crew of only, say, a dozen, would become a realistic proposition. Without it, larger ships and larger on-board communities seem more likely.

Although speeds of this order do open up the possibility of round-trip interstellar missions accomplished within the lives of the crew, and within the lives of the mission controllers, it does not follow that all such missions will be designed to return the astronauts. It may be that there are some who would prefer to live out their remaining days exploring the cosmos rather than returning to the place of their birth. In a sense they would opt out of human society, still returning information to the mission controllers (although, if they decided not to bother, there is little that could be done about it), but content to keep their own company throughout their lonely voyages. How many of us on this overcrowded planet might be tempted by such a prospect?

Relativistic Starships
As we have already seen, at velocities which are a high fraction of the velocity of light, time dilation effects become dramatic. For example, at a steady velocity of 99% of that of light, 100 years of Earth time would correspond to only 14 years of ship time. If we allow 1 year for each acceleration or deceleration period, then a three-parsec return mission with a cruise velocity of $0.99c$ would occupy about 22 years of Earth time, plus the actual exploration period; say 25 years in all. On board, the elapsed time, including exploration, would be about 9 years. The returning crew would find that they had aged by 16 years *less* than the Earthlings they had left behind, but the time difference would not be so

acute that they would not be able to recognize or communicate with family and friends.

On board such a ship a fairly conventional command structure would be quite feasible, since on a three-parsec exploration mission no one on board is going to age by more than 9 years—still a long time to live cheek-by-jowl, but acceptable to some at least.

The prospect opened up by relativistic spacecraft, though, is of travelling vast distances. At 99% of the speed of light a round trip mission to a range of 30 parsecs would occupy approaching 35 years (ship time), including acceleration/exploration time, but more than two centuries would have passed by before the spacecraft returned according to terrestrial clocks. The problems posed by this situation are considerable. The 'ideal' $1g$ spacecraft described earlier allows all possible journeys in our Galaxy to be attainable within a normal human lifespan. A complete circumnavigation of the Galaxy would occupy less than 25 years of ship time if the spacecraft accelerated at $1g$ for half the voyage and decelerated at the same rate for the second half. Some 200,000 years would have elapsed on Earth in the meantime, and there is no telling what the astronaut crew would return to.

The long range relativistic spacecrew can accomplish—in principle— almost any voyage it cares to undertake but it cannot return to its own parent society. By undertaking missions of hundreds or thousands of parsecs, the crew would cut themselves off from Earthly ties just as surely as would the community which embarked on a more stately and pedestrian journey in a Space Ark. It may be that most two-way voyages, even at relativistic speeds, will be confined to distances of less than a few tens of parsecs simply because of temporal effects. If human longevity increases, then longer two-way journeys would become more meaningful and, indeed, increased individual longevity may lead in turn to more stable, long-lived cultures; the pace of social change may diminish. If so, then the returning relativistic astronauts may find it less difficult to re-adapt to life on an Earth now centuries or millennia in advance of its state when they left. Alternatively, the training which they may have received in 'techniques of communicating with aliens' may be essential to enable them to communicate with the new generations of terrestrial (or Solar System) inhabitants.

Perhaps the relativistic travellers will find it necessary to travel in family units so as to be able to ensure the continued existence of their species in whatever remote parts of the Galaxy they finally land and settle. Rather than explorers, whose duty is to report back to Earth, they would become colonists.

Each form of interstellar travel has its problems, and it appears almost as if the slow Ark inhabitant and the long-range relativistic astronaut have similar problems to face in relation to the community 'back home'. Only the fast starship crew, or the short-range relativistic crew, can really hope to return to anything remotely resembling the Earth community they once knew. These discussions, of course, ignore possibilities of 'instantaneous travel' such as we mentioned earlier, for, if such things are feasible, then it is not possible to make any very logical statements about what the effects might be on interstellar travellers, particularly if they are able to arrive before they set out!

Excluding such bizarre possibilities, space—for the long-range astronaut—is likely to remain a lonely place.

32
A SEPS, as seen in the previous plate, sampling the surface of an asteroid by use of sticky-coated nylon ropes weighted with small harpoons. Such a mission could become feasible during the 1980s. (*NASA.*)

33
An artist's impression of a possible modularized space station (three of which are in view) in Earth orbit. A spacecraft is shuttling between the stations, and will dock with the nearby station, as has another spacecraft with one of the more distant. (*NASA.*)

overleaf
A spinning black hole, more than thirty times the mass of our Sun, makes up a binary system with another massive star, itself of nearly twenty solar masses ; the latter is in the red giant stage of its evolution, and now has a diameter some one hundred times larger than that of the Sun. As a result of this expansion, the fringes of the star are close enough to the black hole for the black hole's gravitational attraction to draw material from the star into a turbulent disk of gas (the accretion disk) which circles round the hole ; some of this material is destined to spiral in and disappear forever within the event horizon. The hot gas in the disk emits X-rays and visible light, so revealing the presence of the black hole.

The star and the black hole are separated from each other by a distance of about two astronomical units (about 300 million kilometres), and revolve about each other in a period of about two months.

The black-hole probe in the foreground has been launched from an orbiting starship some seven astronomical units (about 1000 million kilometres) distant from the binary system. The probe is destined to fall into the spinning hole. Its main body contains an instrument module, with microwave and laser communications systems, and is mounted ahead of a rocket motor which is to be jettisoned after burn-out.

This unit is linked to a second instrument section by a structure containing a long aluminium rod to measure gravitational waves, the mass of the second unit allowing the orientation of the probe to follow the local 'gravity gradient' ; gravitational tidal effects can be measured along the length of the rod.

Due to tidal stresses the probe will break up as it approaches the black hole. The forward section will then release a number of very compact

continued on page 181

10 Back Home It's a Long Wait

What will be the implications for the Earth, and for terrestrial society, of the wave of interstellar exploration and colonization? Perhaps the question should really be addressed to the effects on Solar System society for, by the time that interstellar voyages are under way, Man is likely to have explored and colonized much of the Sun's system.

On the Earth, history records great eras of exploration, such as the period around 1500 which saw the bold ocean voyages of Christopher Columbus, Vasco da Gama, John Cabot and Ferdinand Magellan and, later that century, the circumnavigation of the world by Francis Drake (1577–80). This was followed by the era of colonization, exploitation and utilization of the New World. Or again, the voyages of Captain Cook (1768–79) were followed within a decade by British settlements in Australia. The pattern is clear enough—a period of exploration followed by a period of colonization, utilization and trade, initially between the colony and the colonial power, and then on a wider front. As the colony acquires a more cohesive structure, it begins to assert its individuality and independence, and the usual sequence of events leads to

packages which it is hoped will survive entry through the event horizon. *If,* as theory seems to suggest, a spinning black hole is a link between our Universe and another, then these will emerge into a different universe from our own. Just as the Pioneer 10 spacecraft carried a plaque into interstellar space as our first tangible interstellar message, so the black-hole probe will send Man's first message beyond the confines of our own Universe.

34
The exterior of a possible Bernal-type space habitat. The roughly 10,000 colonists, members of a space-manufacturing complex workforce, would live in homes on the inner surface of a large sphere some 1·5km in circumference, rotating to provide a 'gravity' comparable to that of Earth. (*NASA.*)

the colony taking, or being granted, independence. Thereafter it has often been the case that the influence and significance of the former colonial power has declined. The American War of Independence achieved its goal in 1776 and, since then, that nation has gone from strength to strength, becoming the most powerful in the world today; the colonial power, Great Britain, has declined in proportion.

In near-Earth space, the early exploration phase has ended and we are already well into the application and utilization phase, with communications satellites, weather satellites, and Earth resources satellites now an everyday and quite unexceptional part of life. The military aspect is there, too, but most of the time we on Earth are quite oblivious to the great amount of activity going on just a few hundred kilometres above our heads. It is only when some dramatic event occurs, such as the Soviet 'spy' satellite *Cosmos* 954 accidentally falling to Earth in northern Canada in January 1978 and scattering radioactive fragments around, that we sit up and take note of the fact that we are firmly in the Space Age. If the events recounted in Chapter 5 follow their anticipated course, we should shortly be embarked upon a true colonization phase which will eventually take in the entire Solar System.

Will the Earth retain its preeminence in human affairs after a century or two of space colonization has elapsed? As more and more industrial activity shifts away from the Earth's surface, and as the space population increases, we may reasonably expect the importance of the Earth (the 'colonial power') to diminish. I do not for a moment believe that we shall see 'wars of independence' as colonies try to free themselves from Earthly rule (a space colony at L-5 would be a rather vulnerable target in such a conflict); rather we shall see a natural evolution of the centre of human activity away from the Earth itself, leading to a kind of Solar System Economic Community (writing the words in reverse order—which seems to happen frequently with international organisations—gives us the rather nice CESS). This state of affairs could well be with us by the time the first manned interstellar missions get under way.

From the economic point of view, in order to pay for the building of starships, we must have continuous growth such that the cost of a starship becomes only a tiny fraction of the gross human product (GHP). Our tentative estimate of the cost of building the first unmanned starship—including research and development—works out at about 100 billion US dollars (at 1975 values). While admitting this may be wildly in error, it is equivalent to several percent of the GHP at present; such a one-off expenditure on a purely exploratory mission could not possibly be justified at the present time. However, as we have seen, a quite modest level of economic growth would render such an expenditure acceptable within a century from today.

So far as the funding of interstellar missions is concerned there are two basic approaches. One is to pay for the costs out of the global scientific budget, to fund the mission as something which humanity feels is worth doing, and this must certainly be the case for exploration missions, manned or unmanned, in the early phase of interstellar expansion. The Daedalus mission, for example, could hardly be expected to return the investment in hard economic terms. It would instead be a curiosity-based piece of research, the results of which would greatly enrich the fund of human knowledge and experience, just as the planetary probes of today, the Mariners, Veneras, Voyagers and Vikings, cannot claim to be of direct economic value but represent instead a *cultural* activity. However, few would deny that it was from initial curiosity-based research that the present boom in space exploitation has developed. In time, the results of the planetary missions, and of future interestellar probes, will prove to have been the stepping stones to Man's conquest of the stars.

The funding of such missions would almost certainly be controlled by centralized institutions. Cases would be made for funds, and these assessed by suitably high powered committees, who would decide priorities on how the available 'cake' be distributed ('We are sorry, Professor Schwartz, but your Black Hole Probe cannot take precedence over the terraforming project . . .').

However, there are ways in which the cost of manned interstellar missions need not come out of the CESS budget at all. If the construction of the kind of space colonies described earlier proceeds apace then, following the line of argument advanced regarding satellite solar power stations, it is quite possible for the cost of constructing an L-5 colony to be repaid within about a decade. If we double that time, to allow for the cost of installing an interstellar drive and acquiring the necessary fuel, we still have the possibility of a colony being commissioned as an interstellar Ark, paying back the entire construction costs, and being free, then, to push off into interstellar space with a clean slate. In this sense, then, colonizing missions could be made to pay for themselves right at the outset.

Given such an approach there would seem to be no doubt that the cheapest form of interstellar colonization and exploration, *in purely monetary terms*, would be by self-financing slow interstellar Arks. However, the investment in *time* would be enormous. The terrestrial or Solar System society would surely be anxious to find out about neighbouring planetary systems within three parsecs in a period of time of less than five hundred years. The mission planners and controllers surely would be just as keen to learn the results of 'their mission' as would the space traveller to survive to see his target.

If the temporal investment in a Space Ark approach proves to be unacceptably high, then the Solar System culture is going to have to face up to massive financial investments of a nature which, today, would be regarded as very long-term. A 'fast' starship taking, say, forty-five years for a round-trip exploratory mission to a star at three parsecs' range is most unlikely to repay the investment in financial terms (unless it discovers some priceless commodity), but even if it did one might balk at the capital tied up in the mission. Let us suppose that a fast manned scoutship can be built for 100 billion dollars (1975 values). If this sum is tied up for 45 years, then at 10% compound interest in some safer investment it could have earned something like 7300 billion dollars. Over the span of the mission one is considering an 'investment' of over 7000 billion dollars, if one is thinking in terms of financial return.

These figures should be seen against the likely growth in GHP. Looking back at Table 6 we see that even a modest growth rate of 2% per annum leads to a hundredfold increase in wealth in 233 years (from 1971), while a 5% rate leads to a thousandfold increase in about 140 years (the rapid expansion phase of Solar System colonization may well make the higher figure the more likely). A factor of 100 makes the outlay somewhat less disturbing, in effect bringing down the initial capital outlay to about one billion dollars in today's terms, and the outlay would be much less than 1% of the GHP. By way of comparison, the total cost involved in building the 15 Saturn-5 launch vehicles used in Apollo and post-Apollo missions was about 6 billion dollars. In effect, we would be getting a 'fast' starship capable of a two-way mission over a three-parsec range for an outlay equivalent to a couple of Saturn-5s today.

These figures are, of course, sheer speculation. It could well be that I have overestimated the cost, for a glance back will show how the launching costs into space have tumbled in the first two decades of the Space Age. The same kind of reduction may apply to starships, given new technologies and the experience of building the first few.

I have no doubt that the costs can and will be met by a human society that has passed through the crises of the next few decades, but, for the even remotely foreseeable future, the magnitude of expenditure will be such that interstellar flight can be taken only on the basis of funding across the whole human community. The personal starship of the space-faring jet set must lie a very long way ahead, and it may never come, depending on how human society develops.

If—as seems most likely—the speed of light remains an absolute barrier to speeds of travel and communication, then the Solar System community will have to reconcile itself to long-term planning so far as mission results are concerned. The Daedalus mission to Barnard's Star occupies 50 years of flight time plus another six years for signals to return to Earth. Allowing for the initial development of the probe, seventy to eighty years of effort would be tied up in the mission, and no one person could remain actively involved with it from its inception to its conclusion. Even if relativistic spacecraft appear on the scene, a 99% of light-speed fly-by mission to Proxima Centauri will occupy some nine years between the departure of the craft and arrival of results. A manned round trip would take up more than twice that time. We are already involved in space missions of that sort of duration. For example, Pioneer

11 is confidently expected to transmit data from Saturn in 1979, more than six years after launch, while it is hoped that a Voyager craft will investigate Uranus in 1986, nine years after setting out. Missions of about a decade should present no problem at all to the community of the twenty-first or twenty-second century.

Two developments which would lead to psychological, social and emotional problems for Earth dwellers would be suspended animation and/or relativistic travel. As we have seen in the previous chapter, both of these developments would allow space travellers to make long journeys, perhaps of many tens of parsecs, and return to Earth without having aged by anything like the amount by which their Earth-bound colleagues had aged. Indeed, all the astronauts' original contemporaries might well be dead on the return. A new generation of space scientists would have to be prepared to meet the travellers from the past. A society which undertakes such missions will have to have reached a state of sophistication considerably beyond the present; it will have to accept the idea of long-term goals for the species, and look less and less for immediate results. Society today lacks that kind of corporate 'patience'.

The concept of the inheritance of property will have to be revised when a situation arises in which the father can disappear for a few generations and then return to claim his own estate. The interstellar trader who sets off for a fifteen-parsec round trip trading mission might make over his business to the care of his son on his departure; by the time he returns a century or more later (after perhaps only fifteen years of ship time) he may have considerable difficulty in persuading his great grandson, whom he has never met, to return the business to him (even assuming it has not gone bankrupt or been taken over in the meantime). The relativity of time will surely open a fruitful new field for the lawyers of a couple of centuries' time. It will also pose headaches for the taxman. 'Death duties' could scarcely be levied on the estate of a relativistic astronaut when no one can be sure that he will not return on some future occasion, perhaps centuries hence. Capital transfer taxes could be circumvented by the simple expedient of the relativistic traveller retaining his possessions and *employing* successive generations of his family; an appealing idea.

At a more basic human level, apart from short-range 'fast' and relativistic flights (a few tens of parsecs at most), families are going to have to accept the idea that they will never again meet departing colonists, and that even radio conversations with loved ones will involve turn-round times of years and then decades. Humanity has encountered this situation before. In the early phase of colonization, there was precious little chance that a stay-at-home family would ever see departing relatives and friends again, and when colonies were first established in the New World or the antipodes, the only means of communication was by means of letters carried on sailing ships;

correspondence could easily involve a turn-round time measured in years.*

With long-range relativistic flights, say of hundreds of years' duration, the temporal problems would be most acute. How would the Earth community of the thirty-second century react to the arrival of a relativistic crew from the twenty-third century, people who had direct experience of life in the Solar System a thousand years before? Imagine how it would be if we today could be visited directly by, say, Leonardo da Vinci, William the Conqueror, Julius Caesar or Aristotle. Our views of earlier cultures would be drastically altered!

If interstellar flight becomes commonplace, and these temporal effects (due to suspended animation or relativity) frequently become encountered, then society may adjust in a most interesting way. Today on Earth we are strongly aware of the physical limitations of our planet. Instant communication is now available on a worldwide basis, whether we are observing the Olympic Games by direct satellite trnasmission or picking up telephones in our own homes to dial direct to the other side of the world. Direct commercial transportation by Concorde SST can take us across the Atlantic in three and a half hours; a spacecraft can circumnavigate the globe in an hour and a half. Physical distance on the Earth has become relatively insignificant, and there is no real barrier (other than financial and political ones) to visiting any part of the globe, and meeting people of any nationality. With the colonization of the Solar System telecommunication times may lengthen to a few hours, and physical transportation may occupy days, weeks, months or even longer. Even so, there will still be the communal feeling of belonging to the same *time*, of being contemporaries in the same age. New developments in one community can be communicated almost immediately to another.

Once people from different, and perhaps widely separated, *eras* can come into contact, then we may well develop the lack of respect for *time* which we presently have for distance. The sense of continuity and succession of generations will be weakened, and the man in the street (or whatever the equivalent phrase may be by then) will become aware in a very real sense of what physicists realized with the advent of relativity theory—that there is no such thing as *absolute time* (time which flows at the same uniform rate for all observers whatever their state of motion). On Earth we have ceased to be impressed by physical distance; in the future, the Solar System community may come to be unimpressed by temporal separation. What will categorize an individual will be neither the spatial coordinates nor the temporal coordinates of his birth, but the point in *spacetime* from which he or she originates.

Long-range relativistic flights, in particular the $1g$ missions we have

*Space Arks would to some extent overcome this problem on a personal level since whole families and communities would be setting out together; thousands or even millions of people could be involved.

discussed, will nevertheless pose problems to Solar System society for a long time to come. The circumnavigation of the Galaxy, which occupies 25 years of ship time but 200,000 years of Earth time, surely cannot be one which could be treated in terms of making any kind of long-range plans about the possible return of the crew. The cost of such a starship would be high and, although the crew could continue to transmit data back to the Earth, the data rate would be much diminished due to the time dilation effect. In effect, the amount of information gained by a human crew in 25 years would be received on Earth over a period of 200,000 years. On an ultra-long mission like this, human society could do little more than foot the bill and wish the astronauts well.

To survive and thrive beyond the middle of the twenty-first century, as we remarked at the outset of this book, humanity is going to have to attain a state of zero population growth. The colonization of the Solar System will allow the overcrowding on Earth to be eased and, with increased wealth spread hopefully more evenly around, the material lot of the individual should be much improved. In nations where the material standard of living is high population growth has almost ceased; we would expect the human population in the Solar System to follow suit and grow only slowly. The interstellar colonization effort, therefore, is unlikely to be motivated by a drive simply to gain living space for a growing population; instead it will be driven by more complex motives such as those discussed in Part I. Interstellar colonization will be concerned with spreading the human influence throughout the Galaxy.

It is interesting to speculate on what strategy the Solar System would adopt if colonization were taken to imply the establishment of large growing communities in other planetary systems. Would it be more effective to despatch slow Arks carrying large communities or smaller ships moving at higher velocities but carrying fewer people. Consider two alternative possibilities, a space Ark moving at $0.02c$ carrying 10,000 people, and a fast ship ($0.2c$) carrying 200 people in a state of suspended animation, both headed for planets three parsecs away, the fast ship departing at the same time as the Ark. The Ark would arrive after 500 years with 10,000 inhabitants thoroughly imbued with the doctrine of zero population growth; this community would then have to embark on a policy of controlled population expansion. The 'fast ship', on the other hand, would arrive after 50 years with 200 people who had been briefed directly by the mission planners. If the colonists have a population growth rate similar to the present global figure of 2% per annum then what would be the population of the community 450 years after arriving (i.e., just at the time when the 10,000 person ark reached its target)? Taking a doubling time of 35 years we find that the population will have increased by a factor of 2^{13}; i.e., from the initial 200, there should have stemmed more than 800,000 individuals. If both alternatives were to be viable at the same time, the fast ship might seem to offer the faster means of attaining the greater population.

But if the Solar System community should embark upon a programme of long-range interstellar colonization it would have to realize that it was spawning a child, or children, over which it could exert little or no control. By extending its range, Solar System society would be undermining its control over the human species. While it might be possible to take some kind of action to 'subdue' a recalcitrant colony three parsecs away, little could be done to influence the affairs of communities hundreds or even thousands of parsecs away. As the influence of the old colonial powers on Earth has dwindled, and as the influence of the Earth may well dwindle in relation to the Solar System, so would the influence of the Solar System on the Stellar Community decline with the extension of the latter.

The idea of an Earth-centred or even Sol-centred galactic empire does not really seem a viable proposition. The light-barrier saves us from ever being faced with a single monolithic human culture throughout the Galaxy, and brings instead the prospect of a truer freedom for individual communities, sharing a common origin, but developing in possibly quite different ways. We shall examine the possibility of contact with aliens in the next chapter; but, even if we never encounter other intelligences, by the process of interstellar exploration and colonisation we will generate our own aliens. Genetic changes adapting humans to strange environments, coupled with social and philosophical developments, may well generate communities whose commonality with old *Homo sapiens* may be marginal at best.

Following the road to the stars will lead to an enrichment of human experience and an enhancement of the opportunities open to mankind. The Earth itself should become a much pleasanter place in which to live, and the terrestrials will benefit by proxy from the widening human horizons. However, the inhabitants of Earth and Solar System alike should entertain no delusions about their continuing preeminence— they can no more hope to retain absolute control over the enlarged human sphere of influence than can a parent hope to determine the future course of his or her child's life. The pleasures and the trials of parenthood lie ahead for the community of Mother Earth.

11 Neighbours

The relationships between colony and colony and colonies and the Earth are likely to develop in a variety of interesting ways. As the Solar System community evolves there may emerge some kind of dichotomy between the planet-dwellers and the inhabitants of artificial space colonies. In the latter, the environment would have been designed to suit the inhabitants and could be modified by them by common consent; even the physical location in space of these colonies could be changed if the inhabitants so desired. The physical environment of the space colonists would be restricted in terms of the volume of the colony, but visiting other colonies should not be difficult, and the inhabitants might not feel any more constrained than would city dwellers today. An expedition to Earth, with its relatively wild natural environment, might constitute a great adventure. The planetary colonists would probably face a more rigorous life, either living in closed environments and exploring beyond with the aid of life-support systems, or transforming their world by some kind of terraforming technique. They might well have to rely on orbiting colonies for support in terms of supplying power, zero-gee and high-vacuum technological facilities, and the like. The planet people might be of a more pioneering nature in the traditional sense, struggling against their physical environment, pushing back physical frontiers. In the larger physical space available on a planetary surface there might be more scope for the 'loner', the individualist, to go his own way. The difference between the planetary people and the space colony inhabitants might be likened to the distinction (rather artificial, I feel) between the outdoor type and the city lover.

The planet dwellers would not have the ability to change their environment over a short period of time, nor could they do anything to alter the inherent gravitational attraction of the planet. Colonists on the Moon would have to adapt to one-sixth of terrestrial gravity, those on Mars to two fifths gee, and so on. If the experience of Skylab is anything to go by, the human body will adapt rapidly and will evolve to suit the changed environment; it seems quite likely that dwellers on low-gravity planets might not be able to return to Earth for other than short visits. To a Mars-dweller arriving on Earth it will seem as if his body weight has increased by 150%, and his system may not be able to take it. He would find it much easier to visit a rotating space colony, as by choosing an appropriate level he could feel his normal weight.

189

Inhabitants of space islands may find after a time that it is more convenient to live in a low-gravity, or even zero-gravity, environment, and may adjust the rotation rate of the colony to suit. In that case they, too, would find it difficult to visit the Earth (or other planets) or even to visit other colonies which decide to stick with terrestrial gravity (although there will always be zones of suitable 'gravity' in spinning biospheres).

Physical contact with the Earth and its inhabitants may diminish, relatively speaking, as time goes by and the space community grows, but near-instantaneous telecommunication will remain. The colonists—wherever they may be—will be within a few hours' communication time of the Earth or of other colonies. Provided they do not elect to cut off communication (and a space colony would be the ideal place for a closed religious order) the colonists would be fully aware of progress and developments elsewhere in the Solar System. There would still be, then, some kind of federated human community within the Solar System and trade and two-way communication, including physical travel between colonies, would take place on a large scale.

On the other hand, it should be possible to reverse, or at least diminish, the current tendency on the Earth towards one single global culture. A single world state has much to commend it, but it will be gained at a cost—the loss of a fair measure of individuality and the submergence of a wide variety of separate cultures. The colonies will offer a new opportunity for diversification without the same kind of confrontation that presently arises where two opposing ideologies meet. Colonies in conflict can back down in a way that opposing neighbour states on Earth simply cannot do; they can physically move apart. In any case, there is no reason for one colony to wish to capture and take over another colony when without too much effort and without risk of retaliation or defeat they can simply build another—but perhaps I am being very naïve about human nature in making such a remark.

Nevertheless, I am inclined to think that relations between human colonies, Earth-bound, planet-bound, and sky-borne, will be amiable, and that the colonization of the Solar System will allow diversification to take place in harmony. In this scheme of things, as we have already discussed, the Earth is unlikely to remain the dominant controlling influence.

Colonies established in other star systems inevitably must be more isolated than Solar System colonies because of the vastly greater distances and communication times involved (even Proxima Centauri is about a quarter of a billion times further away than the Sun). Such colonies must be totally self-reliant and independent of the Earth. If located less than ten parsecs away the colonists would be able to sustain some kind of dialogue with their parent planet and keep in touch reasonably with the trend of developments there; likewise the Earth would be able to keep reasonable track of the growth, successes and failures of the colonies. Even so, effective *control* of the colonies by the Earth would hardly be possible; imagine the difficulties of a military commander whose information about the outcome of a battle is ten years out of date, and whose response takes a further ten years to reach the scene!

As the zone of colonization spreads to hundreds or thousands of parsecs' range the lines of communication will become more severely stretched and colonies are likely to establish stronger relations with their nearer neighbours than with the Earth. Each will be transmitting and receiving information, but because of the differing signal transit times no two stellar colonies can be in possession of precisely the same body of knowledge. On Earth it is, in principle, possible for all libraries to maintain an identical stock of information because communication is instantaneous. Within the Solar System community, for all practical purposes, the same would hold true. Beyond that the communication time will ensure that uniformity is not possible. The Universe seems to be constructed in such a way as to ensure the existence of a wide variety of communities and, therefore, to enhance the survival prospects for the phenomenon of intelligence. The collapse of a single, interdependent, Galaxy-wide civilization would be a disaster, for it would drag down the whole intelligent community. As the light-barrier precludes the possibility of a strongly interdependent galactic community, the collapse of individual colonies would have no major effect on intelligence as a whole. It is a basic principle of ecology that, the more complex an ecosystem, the less severely it is affected by individual events and the more chance it has of surviving change; the same most probably will be true of galactic society.

This argument does not preclude the possibility of social interaction, trade and, possibly, conflict between neighbouring colonies. The strategy adopted for the colonization process may well give rise to the possibility of conflict. Will all starships be launched from the Earth, or is it more likely that starships will establish a number of widely separated colonies, each allocated a zone which they are to colonize by building their own starships? For the sake of argument, let us suppose that the Earth despatches six interstellar Arks to six colonization centres six parsecs away from the Earth and from each other (fig. 25) and let us suppose each is given the task of colonizing all star systems within a three-parsec radius of their site. The craft to be used is the trusty 10,000-person Ark, having a cruise velocity of 0.02c.

In our region of the Galaxy each zone would contain seven to ten stars (let us assume ten for simplicity). The initial outward flight would take 1000 years. Allowing 50 years for the community to establish its base and assuming thereafter a population doubling every 35 years, we may approximate that suitably populated starships should be despatched every 35 years to the various zoned stars. If the number of stars is 10, then the launches will be spread over 315 years, and if we also assume that the mission to the most distant star (at about three parsecs' range)

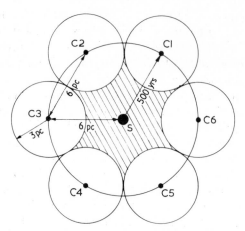

Fig 25 **Colonizing our immediate stellar neighbourhood — one approach.**
Starships (Space Arks travelling at 2% of light speed) are despatched to six
colonization sites at a range of 6 parsecs, each craft taking 1000 years to
reach its target. Thereafter each site colonizes the star systems within a
radius of 3 parsecs; the shaded area is colonized directly from the Solar
System. The whole operation would occupy 1500 to 2000 years.

is undertaken last (an unlikely possibility), then the final starship will
reach the most distant zoned star 500 years after launching. The time
from leaving the Earth to filling the allotted zone is 1000 years plus
50 years plus 315 years plus 500 years—i.e., 1865 years. If the same also
applies to the other five zones then each centre will have expanded to
its allotted limit by this time and will have reached the boundary of the
neighbouring zones. If past experience of the human race is any guide,
territorial conflicts may arise at this stage.

Given a faster drive, say 0.1c the range of colonization centres might
be increased to 30 parsecs, giving each centre a sphere of 15 parsecs'
radius to colonize, within which there would be about 1000 stars. The
time required to fill such a zone would be governed more by the rate of
construction of Arks than by their speed. Thus it would take 500 years
to reach the most distant stars in the zone, but—assuming one starship
launch every 35 years—it would take 35,000 years to despatch enough
starships. If territorial conflict were going to arise, it would do so by
about this time.

It is interesting, as a purely speculative exercise, to make a rough
estimate of the time needed to 'colonize' the entire Galaxy in this sort of
way; i.e., with the Earth despatching starships to colonization centres,
each separated by 30 parsecs, and the colonies thereafter filling in their
zones. If we establish centres in the galactic plane and allow them
each to colonize a 'cylinder' of space above and below the galactic
plane (fig. 26), then we find that several hundred thousand such centres
would be required, and on average each centre would have access to
several hundred thousand stars. The longest journey from Earth would

be of the order of 30,000 parsecs which, at 0.1c, would require one million years but, even so, the time taken to build the starships (at a rate of one per 35 years) would be the dominant factor. 3.5 million years for each hundred thousand starships. Let us suppose it takes 10 million years to launch sufficient starships to the centres (phased so that the longest flights depart quite early in the scheme of things). Each zone would require a similar period of time to build sufficient starships to occupy its zone; but, with appropriate phasing of launches to the centres, the entire operation need occupy only 10 million years or so.

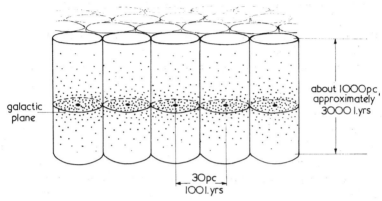

Fig 26 **The colonization of the Galaxy.** If such an onerous task were to be undertaken, one approach might be to establish colonization sites separated by about 30 parsecs in the galactic plane. Each site would then be responsible for colonizing all the star systems vertically above or below its zone; i.e., each site would colonize a cylinder about 15 parsecs in radius and 1000 parsecs long—the thickness of the galactic disc in our area being something like 1000 parsecs.

Of course, these figures are hypothetical in the extreme, and refer to a policy (the complete colonization of the Galaxy) which is scarcely likely ever to be undertaken. Indeed, because of the evolution of diverse societies and different human species, and the communication problem discussed earlier, it would almost certainly prove impossible to maintain the coherence of such a plan (and a very good thing, too). Nevertheless, as a speculative exercise, the figures are interesting in that they would seem to imply that the Galaxy could be colonized in a period of time which is short compared to the age of the Galaxy itself (over 10 billion years). This has some intriguing implications when we come to consider what is perhaps the most important question, the chances of and consequent effects of an encounter with an alien intelligent species.

Aliens
In Chapter 2 we examined the likelihood of life existing elsewhere in the Universe and found that, without making any particularly extreme assumptions, life, in some shape or form, was likely to be abundant in

the Galaxy. By making further quite cautious assumptions about the evolution of intelligence, civilization and technology we concluded that there is a good possibility that there exist other advanced technological societies in the Galaxy at this time, societies capable of communicating over interstellar distances. The figure which arose from our deliberations was 100,000. This of course is entirely speculative, but is more conservative than the values commonly quoted in discussions of this kind. If this value were correct, then the *average* distance between communicative technologies would be of the order of three hundred parsecs, but it is possible by chance that we have one located much closer than this.

How might we ascertain the existence of our alien neighbours and establish communication. Clearly there are several possibilities: *direct contact*, in which they visit us or we visit them—this might occur by accident or design; *long-range contact* by means of radio (or other) transmissions—we may receive their message and send a reply, or they may receive ours and send a reply.

Let us consider radio communication first, since we already have the means at our disposal. The topics of the search for extraterrestrial intelligence (SETI) and communication with extraterrestrial intelligence (CETI) have been the subject of a number of serious conferences, studies and publications, notably Project Cyclops which was discussed earlier (we looked at some of the technicalities of transmitting and receiving over interstellar distances). Possibilities that have been considered include:

(a)   detecting some kind of *beacon*, the assumption being that an advanced technological society would wish to advertize its presence by means of an omnidirectional transmitter radiating some kind of repetitive message;

(b)   *'eavesdropping'*, the assumption being that there already exists two-way communication between more advanced civilizations in the Galaxy, and that we might accidentally intercept such a transmission;

(c)   *'deliberate beaming'*; i.e., we pick up a message deliberately directed towards us, either by a more advanced civilization which knows of our existence, or by one which is probing suitable stars on the off-chance of receiving a reply.

Clearly, for a given power of transmitter, it is easier for us to intercept a directed message (being transmitted in a narrow beam) than an omnidirectional one. We already possess radio telescopes capable of receiving messages transmitted by similar pieces of equipment tens of thousands of parsecs away, provided that we happen to be looking in the right direction and are tuned to the right frequency at precisely the right time. Unless an omnidirectional beacon were of exceptional power, its range would be limited. Nevertheless, Project Cyclops came to the conclusion that, for an outlay less than the cost of the Apollo programme, we could construct an array of radio telescopes (some 5 kilometres in diameter) which could pick up a 1000 Megawatt beacon at a range of 300 parsecs. This transmission power requirement is not extreme, amounting to only 10% to 20% of the power output of one of the proposed satellite solar power stations discussed in Chapter 5. Whether or not an advanced society would *wish* to advertize its presence in so crude and blatant a way is open to question.

Considerable discussion has centred on the appropriate frequencies at which to search. As we mentioned, background noise in the sky is at a minimum in the microwave region of the spectrum (1–60 GHz) and this may be the optimum waveband to choose. At a wavelength of 21 centimetres (i.e., a frequency of 1420MHz), neutral hydrogen gas in the Galaxy emits radiation, and it is argued that any society which developed radio communication would be aware of this and would choose to transmit at this 'natural' calling frequency. This line of argument was advanced in a paper by C. Cocconi and P. Morrison published in *Nature* in 1959, and was taken up by Dr Frank Drake of the National Radio Astronomy Observatory, Greenbank, Virginia, where Project Ozma was initiated in 1960. This was the first recorded attempt to pick up signals from alien intelligences, and concentrated on a listening programme at 1420MHz, the radio telescope being directed towards 'suitable' nearby stars, Epsilon Eridani and Tau Ceti.

No success was reported, and for a time there was little activity in this field, but during the past ten years interest in such searches has grown with several different programmes being initiated, notably at NRAO (Greenbank), Gorky (USSR), Arecibo (Puerto Rico) and the Algonquin Radio Observatory (Canada). Various frequencies have been studied. There is now quite widespread feeling that 1420MHz may *not* be the most appropriate frequency—after all, if hydrogen emits there, it is going to be a noisy frequency! Between the hydrogen emission line at 1420MHz and the OH (hydroxyl) emission line at 1662MHz is a region of particularly low background noise and the Project Cyclops study suggested that between these two limits would be the natural interstellar frequency range. Since H plus OH equals H_2O (water), this region of the microwave spectrum has acquired the delightful title of 'the waterhole'. What more natural place, argues Dr B. M. Oliver, could there be for the meeting of communicative species?

To date there have been no detections of interstellar messages (although the discovery of pulsars did cause a bit of a flurry in 1967/68), but little can be deduced from this. A definitive search programme would require a set-up of the Cyclops type capable of listening to 100 million transmission channels simultaneously; allocating 1000 seconds only to each star, several decades of steady work would have to be allocated to the task of scanning all suitable stars out to 300 parsecs.

We have, of course, taken the first few steps in *transmitting* information about our presence to the stars, in the form of, for example, the 1974 Arecibo transmission to the cluster M13. We have also attached a plaque to the Pioneer 10 vehicle, and included a recording of 'the Sounds of Earth' on the Voyager probes just in case 'someone out there' should happen to find them. None of these approaches is likely to elicit a response in the foreseeable future! Devising a suitable message which is universally decodable is of course an overwhelming task. Even intelligent terrestrials have difficulty in making sense of these messages; goodness knows what a true alien would make of them. Nevertheless, the first steps have been taken.

35
An artist's impression of the interior of the space colony shown in the previous plate. Near to the 1·5km equator wanders a small river whose shores are made of lunar sand; here the artificial gravity is the same as that of Earth, but at polar regions it would be zero, permitting low-gravity sports and travel by human-powered flying machines, one of which is shown. Otherwise travel within the community would be on foot or by bicycle. Heat and light are supplied by natural sunshine beamed inside by the external mirrors shown in the previous plate. (*NASA.*)

36
A model of a proposed manufacturing facility for a space colony. Here material brought from the Moon would be converted into glass, aluminium and other useful materials. The facility would be situated several kilometres from the space colony and connected with it by a transport tube. Power is provided by solar radiation captured by the triangular solar panels shown. (*NASA.*)

37
Two large O'Neill cylinders (32 × 6·4km), a form of space colony suggested by Gerald K. O'Neill of Princeton University, each capable of holding between 200,000 and several million inhabitants, depending upon the planning of the interior. Each cylinder rotates every 114 seconds to provide an artificial gravity the same as that of Earth. The power source is solar energy, and raw materials for construction are mined from the Moon or from the asteroids. The teacup-shaped containers ringing the end of each cylinder are agricultural stations, and at the centre of the rings are manufacturing stations. Long rectangular mirrors at the sides of the cylinders and hinged at their lower end reflect sunlight into the interior, regulate the seasons and control the day—night cycle. It has further been proposed that O'Neill cylinders could be used as generation ships, often termed Space Arks. (*NASA.*)

38
A segment of the colony shown in the following plate while in the final stages of construction; in the interior can be seen an agricultural area with a lake and a river. (*NASA.*)

35▲

36▼

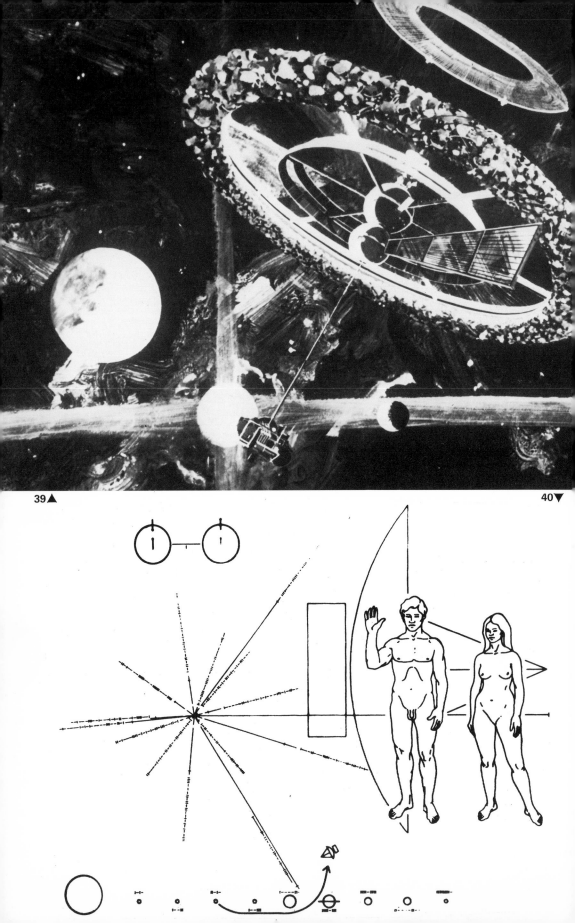

The rationale usually applied to the construction of interstellar messages is the 'universality' of mathematics and the laws of nature. It is supposed that any advanced technological civilisation will have developed the same modes of mathematical expression, and will be aware of the same basic 'laws' by means of which we believe the Universe to be governed. Even here we are probably being over-anthropomorphic (or 'Earth-chauvinist', as Carl Sagan would say), for it is surely quite conceivable for an intelligence, and a culture, to arise which thinks and communicates in a fundamentally different way to ourselves and to which mathematical logic is wholly alien—as it seems to be in any case to the majority of *Homo sapiens*. Nevertheless, it is probably true to say that any civilization capable of *receiving* our message would have a similar technology, and therefore a similar approach, to our own. To receive a radio message presumably requires a radio receiver which must take the same general form as our own devices.

39
A proposed space colony capable of holding some 10,000 people; the wheel-like colony has an external diameter of about 1·5km and an internal diameter of about 200m. The massive exterior of lunar rocks provides shielding from cosmic radiation, and artificial gravity would be obtained by rotating the colony once per minute. The mirror floating above the colony reflects sunlight into the ring mirrors below, which in turn reflect the light through 30m strip windows into the colony's interior to provide light and assist agriculture. Projecting from the lower ball of the colony is the long rectangle of a heat radiator, which dissipates excess heat into space; at the end of a long transit tube projecting beneath the colony is the manufacturing area, as depicted in the previous plate. (*NASA.*)

40
The plaque carried by the first manmade craft to follow the road to the stars, Pioneer 10. In the remote chance that some scientifically educated extraterrestrial civilization should come across the craft, millions of years in the future, clues are given as to the origin of Pioneer and the type of beings who created and launched it. The radiating lines at left indicate the positions of 14 pulsars with respect to the Sun, which lies at the point of intersection of the lines; the '1' symbols along the lines give the frequencies of the pulsars (in binary code) relative to that of the hydrogen atom (shown at the upper left with a '1' symbol), so that by locating the 14 pulsars using this universal clock the extraterrestrials should be able to pinpoint the Sun and, by calculating the rate of decrease of the frequencies of the pulsars, to determine the time that has elapsed since Pioneer was launched. The human figures are given scale both by the outline of the spacecraft and by the binary number just to the right of the woman's figure; the binary number is equivalent to our decimal 8, and multiplication of this by the wavelength of hydrogen (about 20cm) gives her height as 1·6m. (*NASA.*)

The usual method of making up a message for transmission is to use the binary code (see page 125) to assemble information which may be decoded into an elementary picture. For example, the message to M13 consisted of a set of 1679 bits (each of which was a 'zero' or a 'one'). 1679 is the product of two prime numbers, 23 and 73 (a prime number being a number which cannot be obtained by multiplying together two whole numbers; thus '17' is prime whereas '18' is not, for $18 = 3 \times 6$ or 9×2), and it is assumed that the receiving civilization will recognize this fact. The information can be set out as 23 columns of 73 rows, or as 73 columns of 23 rows. Adopting the former strategy, we can draw up a grid in which each little square contains either a 0 or a 1, and if we chose to shade in the squares containing 'ones' and leave blank the squares containing 'zeros' we will end up with a picture which looks like some kind of demented crossword puzzle, but which contains a surprising amount of information. The message includes the following information: the numbers 1 to 10; the atomic numbers for the elements hydrogen, carbon, nitrogen, oxygen and phosphorus; the formulae for the basic structure of human DNA; the basic form and scale of a human being; details of the Solar System; and the scale of the transmitting antenna. The unit of size employed is the wavelength of the transmission itself.

What I have described is the information the message was *intended* to convey; there is no guarantee that the recipient—if he makes anything of it at all—will not place a wholly different interpretation on it. The almost total lack of real communication between individuals—and political groups—on the Earth points up the difficulty of communicating with aliens. The problems are further highlighted by the existence on Earth of creatures which are evidently highly intelligent but with whom effective communication has so far proved impossible. I am thinking, of course, of whales, dolphins and their like. No, establishing communication with another species, whether remotely or in person, will prove to be a task of the utmost difficulty; it really would be unrealistic to expect the archetypal alien to step from his craft with a fixed grin and a halting rendition of that classic phrase, 'Take me to your leaders.'

The chances of direct contact are hard to assess. If we assume, purely for the sake of argument, that the nearest 'advanced' civilization lies at a range of 300 parsecs, then to reach that distance would take 50,000 years in a 0.02c Space Ark, 10,000 years an a 'fast' (0.1c) craft, and about 1000 years at the speed of light. These times assume that the mission is specifically directed towards the correct target system. Random searching would take much longer. If the 'colonization' of the whole Galaxy were undertaken along the lines described earlier (using fleets of Arks) then presumably if there is any long-lived civilization elsewhere in the Galaxy we inevitably must encounter it by the time the operation is complete (a possible timescale of 10 million years). On the other hand, our nearest neighbour might be only a few tens of parsecs away, and we might make the direct encounter within a couple of centuries. To put a timescale on this sort of encounter is to resort to pure guesswork at this stage of our development.

It is, of course, possible that the encounter may come the other way, that they may visit us. In fact, if 'they', meaning advanced technological civilizations, do exist this is the more likely course of events. Suppose that there are, for the sake of argument, 100,000 technological civilizations in the Galaxy at present; in order to come up with that number (page oo) we had to assume a mean lifetime of a technological civilization of something like 100 million years. We have been 'advanced' in the sense of possessing radio communication for less than one hundred years (Marconi sent his first transmission in 1895). These hundred thousand civilizations will be spread out in 'advancement' between a stage we reached just at the turn of the century, and a hundred million years 'ahead'. On that reckoning it is quite likely that we are the only one which lies within one thousand years (either way) of the invention of radio. We may be, for the moment, the *least* advanced communicative civilization in the entire Galaxy.

There may be other cultures in the Galaxy who have possessed the means of space travel for millions of years—even, perhaps, billions of years. If space-faring civilisations are common in the Galaxy then we ought to expect to be visited before we ourselves do any visiting. There is no acceptable evidence to show that we have been visited in the past, or are being visited at present, by extraterrestrials. There are those who maintain that both of these suggestions are true, but the 'evidence' produced is usually open to a multitude of interpretations. If either contention were proved to be valid, the discovery would be the most significant ever to be made; its importance could not be overestimated, and the philosophical, social and scientific impact would be profound. Such a discovery would demand evidence of the highest quality and integrity; the evidence is lacking at the present time.

Unidentified flying objects (UFOs) are often cited as evidence that we are being visited by alien spacecraft. I do not deny that UFOs exist in the sense that people see phenomena in the sky which behave in an unfamiliar way, and which cannot be identified. Most sightings are subsequently answered by some very ordinary rational explanation but it must be admitted that, even after serious scientific investigation, there remains a residue of UFO reports which cannot be explained away. To insist that these must be alien spacecraft, however, is absurd in the extreme. There remain many natural phenomena of which we know little and understand less. The UFO phenomenon merits serious investigation and its explanation may invoke physical and psychological factors. *It is not impossible* that we have been or are being visited by alien spacecraft, but it is irrational to demand that the only explanation of all or even some of the sightings is this one. There is, incidentally, an interesting circular argument associated with this 'explanation'. If advanced technologies are relatively common in the Galaxy, then the emergence of life, and intelligence, must be commonplace also; if so, there is nothing unusual about us, so why should we be the particular focus of all this space activity? On the other hand, if intelligence is rare, and space-faring technologies very scarce, it is surely odd that so many 'spacecraft' should visit the Earth out of the multitude of planets in the Galaxy.

The 'alien spacecraft' explanation of UFOs says more about the apparent psychological need of much of humanity to believe in some higher power than it tells us about the prevalence of advanced technologies in space. Make no mistake, it is possible that tomorrow or the day after an alien starship will settle into 'standard orbit' around the Earth, and it is possible that another culture has explored the length and breadth of the Galaxy in the past, perhaps visiting the Earth before animal life evolved. The fact which remains is that there is no remotely convincing evidence that aliens have visited this planet.

In considering the chances of meeting up with aliens in the Universe, it is worth considering the two pieces of negative data that we have, that we have not received a message from the stars, and that there is no evidence of visitation by aliens.* The explanation which we offer to account for this may have profound implications for ourselves.

If we ignore the (very real) possibility that the absence of evidence of aliens is simply due to our not having looked hard enough (and this certainly applies to radio searches), we can explain the absence of extra-terrestrials on or around the Earth in a wide variety of ways, some of which are listed below:

(a)   we are unique or, at least, we are the first advanced technological society with the capacity to undertake spaceflight;
(b)   we have been visited in the past;
(c)   there has not been sufficient time for alien spacecraft to reach us;
(d)   there are social and/or psychological barriers to interstellar travel, or to the process of colonization;
(e)   there are physical objections to the colonization of the Earth;
(f)   the 'zoo' hypothesis.

(a) *We are unique.* This is the simplest explanation, but by no means the most acceptable one. Our observations indicate that there is nothing unique about the Sun, and it appears likely that planetary systems are commonplace, so that there is no good reason for assuming the Earth

---

*It may be argued that absence of evidence is not evidence of absence. In an article in the quarterly *Journal of the Royal Astronomical Society* (vol 16, pp 128–135, 1975), M. H. Hart takes as fact ('Fact A') that there are no extraterrestrials on Earth. As L. J. Cox points out in a subsequent number (vol 17, pp 201–208, 1976), 'Fact A' is not a fact, but a hypothesis. The *fact* is that no intelligent beings from outer space are *observed* on Earth now. There is no way that you can be absolutely sure that 'they' are not here cunningly disguised as humans! Surely a culture millions of years in advance of our own could have devised the means to keep an eye on us without our being aware of the fact. Nevertheless, we will assume that 'Fact A' is true in the absence of contrary evidence.

is unique either. Why, therefore, should *we* be unique? The lesson of history is that every time we assigned ourselves particular significance in the Universe we had to revise our opinion later. That we are the *first* is perhaps more plausible but, even so, there are many stars far older than our Sun, and presumably many older planetary systems on which life could have developed long before it emerged on Earth; it would be surprising if none had advanced further than ourselves.

(*b*) *We have been visited in the past.* Insufficient data! We might, however, expect to find artefacts somewhere in the Solar System.

(*c*) *Insufficient time.* We have discussed the possibility that the Galaxy could be colonized within ten million years by a multitude of slow-moving Space Arks, initially despatched from Earth, but also from the selected colonization centres. With an over-riding importance attached to this process, the time could be reduced, perhaps, to only one million years. However, it has been argued that the process might take far longer due to a number of factors—the finite lifetime of starships, two-way traffic between the parent planet and established colonies, lack of crews, all starships having to come from the parent planet, breakdown of the coherence of the colonization plan, etc. However, any explanation must be applicable to *all* potential space-faring communities. It only needs one successful culture to be able to carry out the colonization exercise.

(*d*) *Physical objections to colonization.* The environment of the Earth may be hostile to an alien species or simply 'unsuitable'. A space-faring species is likely to evolve to suit the environment into which it moves. It may be that the construction of colonies in space, possessing controlled artificial environments, is a necessary prerequisite to interstellar flight, and leads to a species which prefers living in such an environment to braving the hazards of existence on a planetary surface. A planetary surface may not be the most convenient place for an advanced species to live (and, as we are discovering, it is certainly not an appropriate place in which to carry out industrial activity). Perhaps aliens visiting the Solar System would be much more likely to hollow out a few asteroids than attempt to colonize planets.

(*e*) *Social and/or psychological barriers.* It may be that the pressures which drive a civilization to develop the high technology that would be required for interstellar travel also prevent that species from under-taking this exploration. Technology has been developed to alter the environment of Man to suit his own requirements (these requirements, of course, are varied, ranging from the basic necessities of adequate nourishment and a roof over one's head to the acquisition of wealth and power on an ever-increasing scale). Its development has contributed to the rise in population and to the depletion of natural resources, to pollution of the environment, and the means of mass destruction. The efficiency of modern communications systems have led to increasing interdependence of the whole of humanity and the consequent vulner-ability of the whole community to a single collapse.

We have already discussed how overpopulation is the major problem facing the world today. The escalating demands for energy and non-renewable natural resources must—if not checked—lead to chronic shortage, and so to the collapse of civilization as we know it. Starvation is not the only consequence of overpopulation: the social effects of overcrowding are already becoming apparent. The collapse of a society whose exponentially increasing demands are met only by an exponential decrease of the means of satisfying those demands is surely inevitable. Fred Hoyle* and others have speculated on the outcome of such a disaster in the case of the Earth. If the collapse occurs before easily accessible fuels and other natural resources are exhausted, then there is some hope that humanity might be able to recover from the collapse and possibly profit by the experience. The further technology advances, and the higher the population becomes, the more severe the collapse is likely to be. If the collapse occurs *after* all easily accessible resources are fully exploited, it may thereafter be impossible for technology ever to appear on the Earth again. To utilize resources which are difficult of access requires high technology, but to develop high technology we must have access to resources; the circle is complete.

If we miss this chance to reach the stars, we may never have another. The collapse of society would lead to immeasurable human suffering, and with the capacity which we now possess might even lead to the destruction of the entire human race.

The global crisis arises for clear and predictable reasons—the finite nature of the Earth and the escalating demands of its population. If intelligent life originates on planets, and develops technology—as it must, if it is to reach the stars—then it seems almost inevitable that the same kinds of crises will arise. What we are experiencing on the Earth today may be a general phenomenon which affects all emerging technological civilizations. Perhaps no civilization has yet overcome the global crisis, in which case many civilizations may have arisen and perished in the Galaxy; perhaps there are many cultures around which have recovered from collapse but which no longer possess advanced technology and which will never be able to leave their parent planets by their own efforts. Perhaps only a few emerging civilizations succeed in overcoming the crises. Because of the self discipline required, the stable community which results may ultimately, as Von Hoerner suggests, succumb to the final crisis of stagnation. Possibly a community which has all its material needs supplied, which has achieved complete stability and uniformity, will have entered an irreversible state. Is there an analogy with thermodynamics here? May we equate a completely stable and uniform society with a state of high entropy? For a heat engine to function, there must exist a temperature difference so that heat can be taken from a high temperature reservoir and converted into work.

*Fred Hoyle, *Ten Faces of the Universe*, W. H. Freeman, 1977.

Perhaps society needs *differences* in order to function; when all are equal in every respect then no 'work' can be done.

Arthur C. Clarke, in *The Promise of Space*, concludes that 'The challenge of the great spaces between the worlds is a stupendous one; but if we fail to meet it, the story of our race will be drawing to its close. Humanity will be turning its back upon the still untrodden heights and will be descending the long slope that stretches, across a thousand million years of time, down to the shores of the primeval sea.' That route may already have been taken by other civilizations which have overcome their global crisis at the cost of losing the urge to expand their horizons.

In searching the Galaxy, if we ever do, we may find evidence of societies which have destroyed themselves, societies which have recovered from their collapse but which no longer have the means to reach the stars, and societies which, having weathered the storm, have attained a state of supreme inertia. Another possibility is that we may find a culture which has no need of technology, and which may never have developed it, an intelligent species so in harmony with its environment that the need to modify it has never arisen. The analogy with whales is apparent again, for these are undeniably intelligent and superbly adapted to their environment. They were roaming the ocean long before the first men scuttled across the land. For millions of years they have survived without any pressure to modify their environment—until Man came along. The uncontrolled butchering of whales over the past century has led to their virtual extinction in a period of time which, compared to their long history, is but the blink of an eye. The vulnerability of such a society to the subsequent arrival of a more aggressive, technologically orientated species is clearly evident.

There is, however, one very simple sociological reason why we should not be surprised at the absence of extraterrestrials on Earth. If what is happening on Earth today is of general applicability, then any emerging civilization is going to have to attain zero population growth, or approach that state very closely indeed. Having achieved that, and perceived the resultant benefits, there will be no pressure to find more living space for a population which is essentially static, apart, perhaps, from some kind of political decision to 'go forth and multiply'. With this condition achieved, the fundamental motivation for interstellar travel would become curiosity, the need to know more and more about the Universe in which the species lived, the desire to discover the existence of other life-forms and intelligences, and—perhaps—the wish to make contact with other advanced civilizations. In consequence of this exploration, the population might increase to a minor extent, but the physical requirement of 'lebensraum' would be removed.

There would be no need for any alien species to 'colonize' the Earth. On the contrary, an alien species might well wish to avoid contact with us.

(*f*) *The 'Zoo' Hypothesis.* The suggestion is that advanced alien civilizations exist, and know of our existence, but are taking good care to avoid making their presence known or exerting any influence upon us. They may be keeping us under observation, to see how we develop, or indeed to see if we destroy ourselves, but they have a policy of leaving emerging cultures strictly alone to develop in their own way. The Earth, then, may be a kind of 'zoo' in which the animals—us—are observed, but are not aware of this observation. It may be so, and indeed I would hope that if we as a species ever find ourselves in the position where we encounter life-bearing planets, particularly those bearing intelligent life, then we will have learned enough to keep well away and avoid interfering. If genetic variety is essential to the survival of a species in a changing environment, surely variety of species and cultures is essential to the survival and development of the phenomenon of intelligence in the Universe. The crew of the future *Enterprise* will, I trust, desist from 'beaming down' and literally putting their feet in it; instead they might choose to observe from a distance, and perhaps leave some kind of remote monitoring station to report back the development of the emerging intelligence.

Perhaps past celestial visitors have done just this with the Solar System. This is the 'sentinel' hypothesis. The Arthur C. Clarke short story of that title describes just such a possibility; the idea was subsequently developed by Clarke and Kubrik in the film *2001—A Space Odyssey*. There is nothing inherently absurd about this suggestion, except —of course—that the information received by the interested civilization would be rather out of date by the time it arrived. Nevertheless, if the velocity of light is truly an absolute barrier to the speed of travel, the order of events will still be maintained; i.e., the 'listeners' would always know in advance if the emerging civilization had despatched a starship in their direction, for no starship could travel faster than the sentinel's message announcing its departure!

Professor R. N. Bracewell of Stanford University has suggested that there may exist a kind of 'Galactic Club' of advanced civilizations which are in contact with each other and which may be experienced at finding developing communities such as ours and inducting them into the galactic community. If they exist and if they are interested in us—both of which are assumptions—then it seems that they will make no effort to establish contact with us until it is quite clear that we are prepared for the eventuality of that contact. Evidence of that preparedness might well take the form of our deliberate attempts at interstellar communication, and/or our first interstellar probes. I do not believe that any Galactic Club—if it exists—would step in to 'save' an emerging civilization from its own global crisis. That surely is the test of the fitness of the species; if it is capable of overcoming the crisis and developing the means of interstellar communication and travel, then it may be regarded as a suitable candidate to join the Club, while, if it fails, it will either eliminate itself or reach some kind of stable existence where the possibility of interstellar interaction is precluded. In the latter case, the resulting culture would not be suited for membership, and would probably be much happier left in blissful ignorance of the Club's existence.

We cannot expect a higher intelligence to step in and sort out all our problems for us. It is here that the cults of UFOlogy and of visitations by men from the stars may hold real dangers, for they may help foster a quasireligious belief that global catastrophe cannot happen because our 'guardian alien' will arrive in the nick of time to save us from the folly of our ways. Our global crisis is of our own making, and we must solve it ourselves; the solutions must be internal ones, and they must be found soon if we are to become one of the galactic species which 'make it'. Anything which underplays the gravity of our present situation can only be counterproductive.

The Consequences of Contact

Contact may occur in one of three situations. Firstly, a more advanced culture may choose to communicate with us (as we have discussed earlier, we may of course discover the existence of other civilizations by accidentally eavesdropping on some interchange; even so, if contact is to be established, the more advanced culture must still elect to respond to any message which we send as a result of this discovery). The more advanced culture will hold all the aces, and will get in touch—assuming it knows of our existence—when and if it suits its own convenience. Secondly, we may encounter a less advanced civilization. This is unlikely in the immediate future as there is little likelihood of there *being* any less advanced species than ourselves who also possess the means of interstellar communication. If, however, we succeed in reaching the stars then as the centuries and millennia pass by we may stumble across emerging civilizations either as a result of direct exploration or as a result of picking up their primitive transmissions. The third possibility is that, by sheer chance, we may encounter a civilization very close to our own stage of development; the chances of this happening must surely be vanishingly small, but it remains a possibility.

The consequences of contact clearly depend on the circumstances of that contact. By far the most probable situation is that contact will be established with a culture far in advance of our own. Will the benefits of such an encounter outweigh the possible dangers? It is usually assumed that any civilization which is far in advance of our own, and which has developed the means of interstellar travel and communication, must be of a benevolent nature. This view is certainly most plausible, for in order to overcome its global crisis, the species concerned must have come to terms with the sources of conflict which we have at present on the Earth, and having presumably accepted the notion of zero population growth will have no need to colonize in the sense of depriving other species of their territory. The degree of stability required for a galactic civilization surely implies that any belligerent race which did not annihilate itself at the planetary crisis stage would do so shortly afterwards.

This is a cosy view, and one which I am inclined to accept. But we are not in a position to state with certainty that it is *impossible* for an aggressive and belligerent species to achieve the means of interstellar travel (after all, such a description could well be applied to us, and we have hopes of doing so). Suppose that on Earth one nation or power group had the means of attaining the total subjugation of the rest of the global population (I don't know what these means could be, but we can all imagine possibilities—drugs, bugs, or simply mass destruction). Given the technology and unrestricted access to available resources, interstellar travel might become accessible to that group, although the long-term stability of the 'civilization' would be questionable.

Encounter with such a civilization, if any such exist, could prove disastrous for us. In that case we may be 'safe' only because they don't yet know of our existence, they don't regard us as a threat, or they are waiting for us to destroy ourselves. Apprehension of the outcome of contact has led some authorities to counsel strongly against attempts to communicate with extraterrestrials. For example, Professor Zdenek Kopal has expressed the view that if we receive a signal from the sky, 'For God's sake let us not answer.' There is the fear of the effects of encountering a malevolent species; there is also the fear, perhaps, of our no longer being the 'top dog' if we encounter more advanced civilizations. We are used to being the dominant species on Earth, and may not take kindly to a more lowly rôle in the scheme of things.

To desist from attempts to establish contact for either of these reasons seems to me to be short-sighted. Can we really remain content to hide under our stone in the hope that no one, ever, will turn it over? If danger exists in the Universe, surely it is better to go out there and meet it than to wait for it to arrive unannounced on our doorstep. I have already said I am disinclined to believe in this possibility—for one thing, if an overwhelmingly aggressive star-faring species had emerged earlier in the history of the Galaxy, surely it would have swept all before it by now—but it cannot be ruled out.

If this fear is a realistic one, then we all should be rather concerned about the unilateral efforts of scientific groups to send out signals on our behalf. But the prospect of, say, the United Nations being responsible for agreeing the contents of every such message would most likely turn radio-astronomers to other less contentious pursuits, like gardening. In any case, the 'damage' has already been done—and not by the radio-astronomers, for much of the UHF radiation broadcast in the form of television programmes escapes through the atmosphere into space. Admittedly this radiation would be hard to receive and probably impossible to decode at any great distance but the fact remains that the Earth from a distance would appear very *bright* at these frequencies in a way that no natural planet could do; ergo, the existence of some kind of technology on the Earth would be revealed. The large scale development of television transmissions has taken place only in the last two or three decades. Nevertheless, the 'leakage radiation' will already have reached out six or ten parsecs.*

The benefits of contact might include the following: firstly the discovery itself—to know for certain that we are not alone in the Universe would have a profound impact upon us; secondly a wealth of new knowledge; thirdly the discovery of new kinds of technology which might be of direct benefit to us (if, for example, the design for an efficient and workable fusion power system arrived by radio tomorrow morning, there would be good cause to celebrate); fourthly, and perhaps most important, interaction with a culture more experienced than ourselves might lend us new perspectives and new insight into the nature and purpose of existence.

However, we must be cautious, for what might seem at the time to be benefits might prove not to be so in the long run. The sudden influx of vast quantities of information, the instant answers offered to many questions which have puzzled us for ages, the donation 'on a plate' of new forms of technology, might submerge our own spirit of inquiry and destroy the very drive which led us to explore the stars in the first place. If a Galactic Club exists, then its members will be aware of these pitfalls and will take good care to avoid saturating an emerging civilization with material which it cannot handle. The principle of 'respect for individual intelligence' should operate to protect the interests of newcomers to the Club.

On the Earth we have had countless examples of the impact of an advanced civilization on a primitive society. The outcome of the encounter has almost invariably been to the detriment of the primitive culture, leading at least to the diminution of that culture and the loss of its individuality, and at worst to its complete destruction. The interaction between different civilizations has had a somewhat different outcome, in that the dominant society (the 'high technology' one) has not itself escaped change; there has been feedback to the colonial power from its colonies, and the cultural change has not all gone in one direction. Where severe effects on less developed societies occurred, it was often as a result of well intentioned desires to 'convert' or to improve the lot of 'the natives' ('improvements' include, of course, the prudish

---

*Freeman J. Dyson has argued that an advanced, high-technology civilization—having a concommitantly high energy requirement—could not hide the effects of its technology. One scheme which he discusses in principle (see F. J. Dyson, 'The Search for Extraterrestrial Technology' in *Perspective in Modern Physics*, edited by R. E. Marshak, Wiley Interscience, New York, 1967) is that a civilization might choose to capture the entire radiation output of its parent star by surrounding it with a sphere made from the materials of a demolished planet. The waste heat radiated from the outside of this sphere would be visible to astronomers as a source of infrared radiation. Nothing could be done to hide this output from the 'Dyson Sphere'. The idea is further discussed by Adrian Berry in *The Next Ten Thousand Years*, Jonathan Cape, 1974.

insistence that they wear clothes to cover their nakedness). We are only now beginning to learn that it is essential to preserve the rich variety of culture on this Earth. It would be ironic if all of that were wiped out by contact with an overwhelmingly superior and dominant extraterrestrial culture.

If we meet more advanced civilizations or a Galactic Club in the Galaxy, I feel sure that when they decide we are worth replying to, then *interaction* will be the key word; they will be as much interested in the new, if limited, perspectives which we may have to offer, as we will be in what they have to offer. 'Domination' will not be the intent of the Club, and in any case, with communication times of perhaps thousands of years, it is hard to see how effective control of one species by another could be maintained.

The encounter with other, or higher, civilizations will set limits on the directions, and ways, in which we ourselves can expand. The physical colonization of the entire Galaxy could not then be regarded as an acceptable aim for us, but, having accepted the notion of zero population growth, that need not be much of a loss. Exploration and interaction with other cultures seems a much more laudable aim for star-faring Man in the future.

The boot, of course, may be on the other foot. If we are the only advanced technological society in the Galaxy at the present time, or if we are the most advanced—at least in our locality—then if encounter is made we will be in the dominant position, and to us must fall the responsibility for the effects on the lower civilization of the encounter. I hope, if doubt exists as to the outcome, we will keep well clear and not make our presence known. We should treat other cultures with the care that we would hope more advanced societies would display towards us.

Should we happen by some remote quirk of fate to encounter a star-faring civilization at the same stage of development as ourselves then, and only then, would there be any likelihood of some kind of conflict. If old ideas of territorial sovereignty and the security of borders still persist, then each could regard the other as a threat. The outcome of a head-on meeting cannot be predicted; all I can say is that such a situation seems utterly improbable. If there are 100,000 technological civilizations in the Galaxy it is most unlikely that even one is within one century of our current stage of development, and to suggest that such an improbable civilization should also be our nearest neighbour is stretching credulity too far.

If all our arguments concerning the probability of life and intelligent life in the Universe are wrong, if we are the first and only intelligent species to develop in the Galaxy, then a great responsibility rests with us. If we believe the phenomenon of intelligence to be important, then we cannot risk restricting it to one habitat (the Earth) vulnerable to a single disaster. Even the Solar System cannot be regarded as sufficiently secure. Admittedly, the colonization of the Solar System would ensure that the destruction of the Earth did not annihilate intelligence, but even the system is not invulnerable to some unforeseen kind of cosmic accident, and in any case in about five or six billion years' time the Sun itself will die. The continued existence of intelligence can be secured only by its inhabiting the widest range of habitats. For this reason alone we must follow the road to the stars.

Face to Face

How will we react to a direct meeting with an alien—not a long-range exchange of information, but a face-to-face encounter? Would there be a reaction of horror and revulsion to an alien whose appearance was, to our eyes, grotesque? Most of us have an unreasoning horror of something, spiders being a case in point; we *know* that spiders—with very few exceptions—are both harmless and useful to us, but how many of us really feel happy about being in close company with them? There is no apparent logic about this dislike, but it is there nonetheless. Would the appearance of a non-humanoid alien affect us in the same way?

We cannot tell, of course. Most of the horrendous creatures of our nightmares are distortions or extensions of familiar everyday creatures. None of our fantasies can prepare us for the wholly unknown. A school of thought suggests that space-faring creatures may well be bipeds, with eyes and other sensory organs on a head at the top of the structure where their range would be greatest; that two eyes would be essential for binocular vision, two ears (or their equivalent) for stereo hearing, and so on. The argument is that, if creatures have evolved on Earth-like planets, they must evolve in a similar way to ourselves because that is the logical way to evolve. The argument may well be valid, but it does seem rather Ptolemaic. It is one thing to see the underlying logic in the structure of the humanoid, it is quite another to state that this same logic must apply in all cases. Evolution, by natural or artificial means, to suit a space environment might in any case alter a humanoid to an unrecognizable state.

An alien may find us just as revolting to contemplate, but at least if we make contact with a more advanced species we may expect that they have been through the trauma of encounter before. A highly developed intelligence might have no need of the sort of organic body with which we are familiar, in which case a 'face-to-face' encounter would be a meaningless concept. The lesson is clear; we can have no preconceptions of the alien which we may one day encounter, the only certain common feature will be intelligence.

If a wholly humanoid alien were to arrive, march up to a policeman in Trafalgar Square, Times Square, or Red Square, and calmly announce (in perfect English or Russian) 'I am from Tau Ceti. Please take me to your leader' then, apart from an immediate reaction, in London, of 'What city, sir?', the eventual fate of the extraterrestrial ambassador would be to be incarcerated in an institution. The more overt approach of flying in direct in a spacecraft could trigger off a destructive reaction in the present state of global mistrust. If he chose to land, say, in the Soviet Union and was persuaded to confer a massive technological advantage on that nation, then the Western power block would be more than a little put out.

No, I do not believe any alien is going to attempt to make direct contact with us here on Earth. If they exist, and if they know about us, they will wisely wait until we have sorted out our internal problems before allowing such an encounter. Any face-to-face meeting will almost certainly be preceded by a period of cautious long-range interaction.

Will we encounter xenophobia ('a morbid dislike of aliens') or xenophilia ('a passionate liking of aliens'); will we find love or hate among the inhabitants of the Galaxy? The answer to that tantalizing question must await the making of contact.

Epilogue
Timescale — the next
10,000 years

Those who make so bold as to try to predict the future should be careful not to attempt to set a definite timescale to events. Lacking even such elementary caution, I shall attempt to set in context mankind's progress along the road to the stars by means of my own guesses as to when major developments may occur. These 'guesstimates' most likely will turn out to be hopelessly wide of the mark, but at least it should be possible to pick out some general trends.

There are many possible scenarios of the future. Taking a pessimistic (some may say, realistic) view, in the early part of the twenty-first century human civilization may collapse in the face of mounting global crises of various kinds. If the collapse is not too severe, then it may be possible for humanity to return to a similar state of development as the present, but with readily accessible resources already worked out the chances of developing the technology for interstellar travel must be severely reduced. Given a more severe collapse, it may not be possible to return at all to a state of high technology. The most pessimistic view is that mankind will destroy itself in some form of global conflict. An alternative view which we have mentioned in Part I of this book is that zero population growth is established soon, and a carefully balanced state of global equilibrium is established.

In what follows, we shall assume that mankind surmounts the global crises, by halting population growth and limiting his energy and re-source-intensive activities on Earth, and expands his sphere of activity into the space environment. We shall assume, too, that for the various reasons mentioned in Part I—and probably for others as well—the human species will embark upon a programme of interstellar explora-tion. Naturally, the degree of certainty which can be attached to any prediction diminishes rapidly the further into the future one looks; we can be reasonably certain as to what will happen in space during the next decade, we can make reasoned estimates of progress in the next century, but with the next thousand or ten thousand years we are really speculating. For this reason the diagram, *The Next Ten Thousand Years*, is set out on a logarithmic timescale; i.e., successive equal divisions on the chart correspond to progressively longer intervals of time. The diagram commences at the beginning of the nineteen-eighties: the first major division takes in the decade to 1990, the second division takes us on a century to 2080, the third division encompasses the next thousand years, and the final division spans the remainder of the next ten thousand years.

The nineteen-eighties will continue the era of exploration in the Solar System, and will also see considerable developments in near-Earth space. Many of these details are discussed, with likely dates, in Chapter 5, and it would be pointless to repeat all these here. Interplanetary probes in the 'eighties should improve further our knowledge of Venus, Mars, Jupiter and, of course, the Moon, and take in previously unexplored bodies, Saturn, Uranus, comets and asteroids, as well as the interplanetary medium. By the mid-'eighties an automatic rover should be crawling across the rugged surface of Mars and atmospheric probes will have plunged into giant Jupiter.

Nearer home, applications satellites of all kinds will increase in numbers and complexity while men and women will become more at home in the space environment as a result of the deployment of the Shuttle, Spacelab and—no doubt—a new generation of Soviet manned space stations. By the mid-'eighties a substantial manned station for six to twelve crew may be established and by the end of that decade we should see the beginnings of a large space-borne industrial complex. On the propulsion front, the new solar-electric propulsion system (SEPS) should see practical application in, for example, the cometary probe, while the solar sail concept may well see practical testing later in the decade. A resurgence in the development of nuclear rockets is possible. The orbital transfer vehicle, or space 'tug', will be developed initially to manoeuvre payloads between the Shuttle and the desired payload orbits and then for broader orbital transfer duties, and towards the end of the decade, development of the heavy lift launch vehicle (HLLV) should be well advanced. A significant development should be the establishment of an experimental satellite solar power station (SSPS) beaming down power to a receiving station on Earth.

The period from 1990 to about 2100 will be the Solar System 'humanization' era, when mankind explores the system in person, establishes colonies, and exploits the resources of the Solar System. So much will happen in this era that one can hardly begin to itemize the major developments. The decade from 1990 to 2000 should see fly-by missions take in the outermost planets, and orbiters or landers explore the others in more detail. A Mars sample return mission may be undertaken by a remote probe around 1990, prior to a manned Mars mission getting under way about the turn of the century. About that time, too, we may see the launching of a high-speed probe to investigate deep space beyond the orbits of the planets.

Manned flights to the Moon may well be resumed by about 1990 in preparation for the establishment of a manned base there in the mid-'nineties, this being a prerequisite for the construction of human space colonies. The manned Mars landing and return mission should lead to the establishment of a Mars base by about 2010 to 2020. As the twenty-first century rolls on, so should manned expeditions encompass the entire Solar System.

On the propulsion front, a nuclear rocket system will almost certainly be developed for use in missions such as the Mars expedition, while the mass-driver (see Chapter 5) should be operational on the Moon by the end of the twentieth century. This system will itself be capable of propelling the asteroid mining vehicles of the mid-twenty-first-century mineral corporations. By the early part of that century there should be significant development towards a workable nuclear pulse rocket or, perhaps, a fusion rocket (see Chapter 7) and it is feasible that an operational system could be available by about 2020–2030.

This era will see the rapid growth of space-based industry and the associated rapid growth of the human population in space, in large space stations and, perhaps, in the first L–5 colony as early as 2010–2020. Satellite Solar Power Stations should be making a significant contribution to terrestrial power requirements by this period. As the century progresses human colonies will expand in size, in numbers and in terms of their distance from Earth and, by the late part of the period, from 2050 onwards, we *may* see space colonies containing millions of people, and we may already be encountering evolutionary differences between the inhabitants of colonies in space, of the Moon and Mars, and of the Earth. Hopefully, back on Earth the birthrate will decline sharply during this period, reaching replacement rate (zero population growth) by the beginning of the twenty-first century; there may be a carefully monitored decline towards an optimum world population as the century proceeds.

Given the development of the appropriate propulsion system it seems likely that the first 'fast' interstellar probe, capable of exceeding one tenth of the speed of light, could be ready for despatch by about 2030–2040, reaching its target system after forty or fifty years in about 2070 to 2080. By the end of this period we should have the first interstellar probe results and will already be embarked on further probe missions leading us into the era of interstellar exploration. The Solar System community will have devised the first self-propelled colonies, either fabricated in the fashion of the L–5 colonies, or constructed inside hollowed-out asteroids, and there may already be the urge among some of the colonists to launch forth into interstellar space; perhaps the first Space Ark will set out before the year 2100, reaching its target around the year 2600.

The interstellar exploration era would have no boundary, but during its first few centuries there could be little in the way of human 'colonization' of other systems. For that reason, I have labelled the period up to one thousand years hence the 'exploration era' and the following period the 'interstellar humanization era', but it must be recognized that, as far as we can foretell, exploration would continue into the indefinite future.

It is of course impossible to foresee when (or if) a true relativistic propulsion system will be developed, a system capable of attaining well in excess of 50% of the speed of light. But if we accept that 10% of light speed can be attained by the middle of the twenty-first century, then it does not seem unreasonable to suppose that relativistic speeds will be attained by the middle of the twenty-second century; let us say the year 2150 (although, for myself, I feel this is a pessimistic estimate—I would not be surprised if such velocities were attained fifty years earlier). The first relativistic probe could have completed a nominal three-parsec mission by the years 2160–2170. Prior to this development, the maximum range to which any interstellar probe could have travelled (again assuming 0.1c velocity is attained not later than 2050) is about three parsecs. From about 2150, the maximum range may increase at something approaching the speed of light (although questions of communication and spacecraft longevity will arise here). The first manned interstellar journeys to a nearby star system *and back* may commence shortly after the first relativistic probes in the late twenty-second century.

Back on home territory, by the late part of the twenty-first century industrial and economic activity in space should far exceed that on Earth; this must be so if the economic resources are to be available to mount costly long-range space missions. By the end of the twenty-second century, the human population in space may well exceed that of the Earth, the latter having declined to a comfortable two to three billion. The middle of this period could see the completion of terraforming Mars or Venus—if that were deemed a desirable thing to do—and by then the Earth itself could well have become a kind of 'nature park', a very pleasant backwater in which to live.

By the middle and late part of this era, small manned exploration bases may be established in a number of target star systems (perhaps among the asteroids in these systems), serviced by relativistic spacecraft while the first Arks will be approaching their predetermined destinations with a new generation of space-faring humanoids who know little of the Earth and may care even less. Perhaps the relativistic spacecraft will be used to arrange welcoming parties.

As the year 3000 is approached, the situation may be as follows: relativistic probes may have attained distances of the order of 150 parsecs, and manned relativistic starships could be not far behind. If the 1g starship, perhaps powered by the interstellar ramjet, is developed within a century of the relativistic drive (say by 2250AD) then individuals may have embarked on the most colossal journeys with every expectation of surviving to arrive at their targets hundreds or thousands of parsecs away.

Throughout the final era on our chart—the 'interstellar humanization

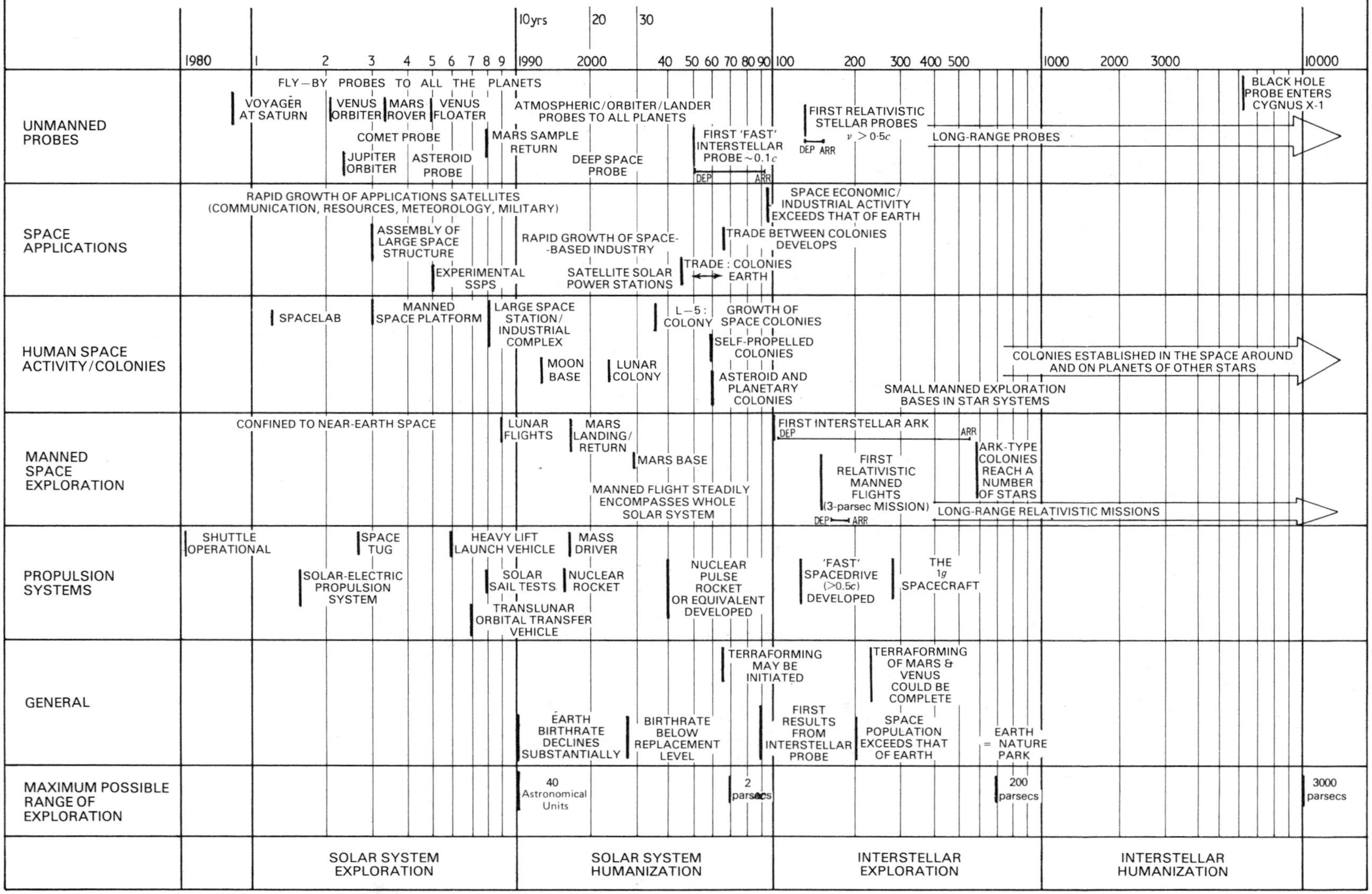

Timeline scale (top): 1980 — 1 — 2 — 3 — 4 — 5 — 6 — 7 — 8 — 9 — 1990 — 10yrs — 20 — 2000 — 30 — 40 — 50 60 70 80 90 100 — 200 — 300 — 400 — 500 — 1000 — 2000 — 3000 — 10000

UNMANNED PROBES

- VOYAGER AT SATURN
- FLY-BY PROBES TO ALL THE PLANETS
- VENUS ORBITER | MARS ROVER | VENUS FLOATER
- COMET PROBE
- JUPITER ORBITER | ASTEROID PROBE
- ATMOSPHERIC/ORBITER/LANDER PROBES TO ALL PLANETS
- MARS SAMPLE RETURN
- DEEP SPACE PROBE
- FIRST RELATIVISTIC STELLAR PROBES v > 0.5c DEP ARR
- LONG-RANGE PROBES
- BLACK HOLE PROBE ENTERS CYGNUS X-1

SPACE APPLICATIONS

- RAPID GROWTH OF APPLICATIONS SATELLITES (COMMUNICATION, RESOURCES, METEOROLOGY, MILITARY)
- ASSEMBLY OF LARGE SPACE STRUCTURE
- RAPID GROWTH OF SPACE-BASED INDUSTRY
- EXPERIMENTAL SSPS
- SATELLITE SOLAR POWER STATIONS
- FIRST 'FAST' INTERSTELLAR PROBE ~0.1c DEP ARR
- SPACE ECONOMIC/INDUSTRIAL ACTIVITY EXCEEDS THAT OF EARTH
- TRADE BETWEEN COLONIES
- TRADE : COLONIES EARTH

HUMAN SPACE ACTIVITY/COLONIES

- SPACELAB
- MANNED SPACE PLATFORM
- LARGE SPACE STATION/INDUSTRIAL COMPLEX
- MOON BASE
- LUNAR COLONY
- L—5: COLONY
- GROWTH OF SPACE COLONIES
- SELF-PROPELLED COLONIES
- ASTEROID AND PLANETARY COLONIES
- SMALL MANNED EXPLORATION BASES IN STAR SYSTEMS
- COLONIES ESTABLISHED IN THE SPACE AROUND AND ON PLANETS OF OTHER STARS

MANNED SPACE EXPLORATION

- CONFINED TO NEAR-EARTH SPACE
- LUNAR FLIGHTS
- MARS LANDING/RETURN
- MARS BASE
- MANNED FLIGHT STEADILY ENCOMPASSES WHOLE SOLAR SYSTEM
- FIRST INTERSTELLAR ARK DEP
- FIRST RELATIVISTIC MANNED FLIGHTS (3-parsec MISSION) DEP ARR
- LONG-RANGE RELATIVISTIC MISSIONS
- ARK-TYPE COLONIES REACH A NUMBER OF STARS ARR

PROPULSION SYSTEMS

- SHUTTLE OPERATIONAL
- SPACE TUG
- SOLAR-ELECTRIC PROPULSION SYSTEM
- HEAVY LIFT LAUNCH VEHICLE
- SOLAR SAIL TESTS
- TRANSLUNAR ORBITAL TRANSFER VEHICLE
- MASS DRIVER
- NUCLEAR ROCKET
- NUCLEAR PULSE ROCKET OR EQUIVALENT DEVELOPED
- 'FAST' SPACEDRIVE (>0.5c) DEVELOPED
- THE 1g SPACECRAFT

GENERAL

- EARTH BIRTHRATE DECLINES SUBSTANTIALLY
- BIRTHRATE BELOW REPLACEMENT LEVEL
- TERRAFORMING MAY BE INITIATED
- FIRST RESULTS FROM INTERSTELLAR PROBE
- TERRAFORMING OF MARS & VENUS COULD BE COMPLETE
- SPACE POPULATION EXCEEDS THAT OF EARTH
- EARTH = NATURE PARK

MAXIMUM POSSIBLE RANGE OF EXPLORATION

- 40 Astronomical Units
- 2 parsecs
- 200 parsecs
- 3000 parsecs

Bottom phase labels:
SOLAR SYSTEM EXPLORATION | SOLAR SYSTEM HUMANIZATION | INTERSTELLAR EXPLORATION | INTERSTELLAR HUMANIZATION

era'—intelligent beings of human origin will spread far and wide; the maximum possible range to which individual spacecraft could have travelled by the end of our chart being just short of 3000 parsecs (i.e., 10,000 light years), and colonies may have been established hundreds or thousands of parsecs from the Earth. To talk about dates and figures here is quite meaningless. The expanded human community will have evolved in a wide variety of ways, and no doubt communication will be maintained between widely separated colonies—but the long communication times will prevent effective control of the enlarged community by any one centre, and the kind of social interaction which results from the continual exchange of out-of-date information will be fascinating, to say the least.

Before 10,000 years have elapsed we may have established communication with other advanced species in the Galaxy. Throughout the foreseeable future a search will be maintained for communications transmitted by alien civilizations—such a message may even be picked up tomorrow. But if intelligent species are separated on average by 300 parsecs then no reply of ours would reach them until the year 3000, by which time we should already be well on the way to attaining that sort of range with actual starships. If more advanced neighbours are awaiting our signal before communicating with us—and if they are 300 parsecs away—then it will be the year 4000 before their reply to our first messages reaches us. In either case there would be time to exchange several communications before the next 10,000 years are up.

Within a thousand years, it is possible that we may meet other advanced species; within ten thousand years there must be a significant chance of having made this encounter if life and intelligent life are commonplace in the Universe. If we are alone then the only aliens we may meet will be our own descendents. At least we should have spread the phenomenon of intelligence far through our local region of the Galaxy, to a maximum range of nearly 3000 parsecs, and into a fully explored region of perhaps 30 parsecs.

I have made my speculations and can now sit back to see how many are wrong. The comforting thing about the long-range ones is that I shall not be here to see how far wrong they are . . . but it would be rather nice to know.

Index

Numbers in italic type indicate pages on which relevant illustrations appear